ASSESSMENT OF WINGTIP MODIFICATIONS TO INCREASE THE FUEL EFFICIENCY OF AIR FORCE AIRCRAFT

Committee on Assessment of Aircraft Winglets for Large Aircraft Fuel Efficiency

Air Force Studies Board

Division on Engineering and Physical Sciences

NATIONAL RESEARCH COUNCIL
OF THE NATIONAL ACADEMIES

THE NATIONAL ACADEMIES PRESS
Washington, D.C.
www.nap.edu

THE NATIONAL ACADEMIES PRESS 500 Fifth Street, N.W. Washington, DC 20001

NOTICE: The project that is the subject of this report was approved by the Governing Board of the National Research Council, whose members are drawn from the councils of the National Academy of Sciences, the National Academy of Engineering, and the Institute of Medicine. The members of the committee responsible for the report were chosen for their special competences and with regard for appropriate balance.

This is a report of work supported by Grant F49620-01-1-0269 between the U.S. Air Force and the National Academy of Sciences. Any opinions, findings, conclusions, or recommendations expressed in this publication are those of the author(s) and do not necessarily reflect the views of the organizations or agencies that provided support for the project.

International Standard Book Number-13 978-0-309-10497-5
International Standard Book Number-10 0-309-10497-1

Limited copies are available from:

Air Force Studies Board
National Research Council
500 Fifth Street, N.W.
Washington, DC 20001
(202) 334-3111

Additional copies are available from:

The National Academies Press
Box 285
500 Fifth Street, N.W.
Washington, DC 20055
(800) 624-6242 or (202) 334-3313
(in the Washington Metropolitan Area)
http://www.nap.edu

Printed in the United States of America

THE NATIONAL ACADEMIES

Advisers to the Nation on Science, Engineering, and Medicine

The **National Academy of Sciences** is a private, nonprofit, self-perpetuating society of distinguished scholars engaged in scientific and engineering research, dedicated to the furtherance of science and technology and to their use for the general welfare. Upon the authority of the charter granted to it by the Congress in 1863, the Academy has a mandate that requires it to advise the federal government on scientific and technical matters. Dr. Ralph J. Cicerone is president of the National Academy of Sciences.

The **National Academy of Engineering** was established in 1964, under the charter of the National Academy of Sciences, as a parallel organization of outstanding engineers. It is autonomous in its administration and in the selection of its members, sharing with the National Academy of Sciences the responsibility for advising the federal government. The National Academy of Engineering also sponsors engineering programs aimed at meeting national needs, encourages education and research, and recognizes the superior achievements of engineers. Dr. Charles M. Vest is president of the National Academy of Engineering.

The **Institute of Medicine** was established in 1970 by the National Academy of Sciences to secure the services of eminent members of appropriate professions in the examination of policy matters pertaining to the health of the public. The Institute acts under the responsibility given to the National Academy of Sciences by its congressional charter to be an adviser to the federal government and, upon its own initiative, to identify issues of medical care, research, and education. Dr. Harvey V. Fineberg is president of the Institute of Medicine.

The **National Research Council** was organized by the National Academy of Sciences in 1916 to associate the broad community of science and technology with the Academy's purposes of furthering knowledge and advising the federal government. Functioning in accordance with general policies determined by the Academy, the Council has become the principal operating agency of both the National Academy of Sciences and the National Academy of Engineering in providing services to the government, the public, and the scientific and engineering communities. The Council is administered jointly by both Academies and the Institute of Medicine. Dr. Ralph J. Cicerone and Dr. Charles M. Vest are chair and vice chair, respectively, of the National Research Council.

www.national-academies.org

COMMITTEE ON ASSESSMENT OF AIRCRAFT WINGLETS FOR LARGE AIRCRAFT FUEL EFFICIENCY

KENNETH E. EICKMANN, *Chair*, U.S. Air Force (retired), Independent Consultant, Austin, Texas
NATALIE W. CRAWFORD, NAE,[1] *Vice Chair*, RAND Corporation, Santa Monica, California
MARK I. GOLDHAMMER, Boeing Commercial Airplanes, Seattle, Washington
STEPHEN JUSTICE, Lockheed Martin Aeronautics Company, Palmdale, California
CLYDE KIZER, Independent Consultant, Nokesville, Virginia
ILAN KROO, NAE,[1] Stanford University, Stanford, California
ELI RESHOTKO, NAE,[1] Case Western Reserve University (emeritus), Denver, Colorado
RAYMOND VALEIKA, Independent Consultant, Powder Springs, Georgia

Staff

MARTA VORNBROCK, Study Director
GREGORY EYRING, Senior Program Officer
JAMES C. GARCIA, Senior Program Officer
DETRA BODRICK-SHORTER, Administrative Coordinator
LaSHAWN SIDBURY, Program Associate (until March 2007)

[1]Member, National Academy of Engineering.

AIR FORCE STUDIES BOARD

LAWRENCE J. DELANEY, *Chair*, Titan Corporation (retired)
TAYLOR W. LAWRENCE, *Vice Chair*, Raytheon Company
FRANK J. CAPPUCCIO, Lockheed Martin Corporation
THOMAS DARCY, EADS North America Defense Company
STEVEN D. DORFMAN, NAE,[1] Hughes Electronics (retired)
PAMELA A. DREW, Boeing Integrated Defense Systems
KENNETH E. EICKMANN, U.S. Air Force (retired)
JOHN V. FARR, Stevens Institute of Technology
RAND H. FISHER, Titan Corporation
JACQUELINE GISH, Northrop Grumman
KENNETH C. HALL, Duke University
WESLEY L. HARRIS, NAE,[1] Massachusetts Institute of Technology
LESLIE KENNE, LK Associates
DONALD J. KUTYNA, U.S. Air Force (retired)
GREGORY S. MARTIN, GS Martin Consulting
DEBASIS MITRA, NAE,[1] Bell Laboratories
CHANDRA N. KUMAR PATEL, NAE/NAS,[1,2] University of California
ROBERT F. RAGGIO, Dayton Aerospace, Inc.
GENE W. RAY, GMT Ventures
LOURDES SALAMANCA-RIBA, University of Maryland
MARVIN R. SAMBUR, Headquarters, U.S. Air Force (retired)
LYLE H. SCHWARTZ, NAE,[1] Air Force Office of Scientific Research (retired)
EUGENE L. TATTINI, Jet Propulsion Laboratory

Staff

MICHAEL A. CLARKE, Director
GREGORY EYRING, Senior Program Officer
JAMES C. GARCIA, Senior Program Officer
DANIEL E.J. TALMAGE, JR., Program Officer
CARTER W. FORD, Associate Program Officer
MARTA VORNBROCK, Associate Program Officer
DETRA BODRICK-SHORTER, Administrative Coordinator

---

[1]Member, National Academy of Engineering.
[2]Member, National Academy of Sciences.

Preface

At the request of the U.S. Air Force, and in light of greatly increased government emphasis on the need for greater fuel efficiency in the fleet of military aircraft, the National Research Council (NRC) was asked to study whether business cases could be made for modifying engines or re-engining large Air Force aircraft. The Committee on Analysis of Air Force Engine Efficiency Improvement Options for Large Non-fighter Aircraft was formed and its report[1] was provided to the Air Force on January 31, 2007.

While that study was under way, congressional interest in fuel efficiency increased, resulting in the inclusion of the following language in Report 109-452 of the House Armed Services Committee on H.R. 5122 (National Defense Authorization Act for FY07):

> The committee commends the Air Force in its efforts to increase aircraft fuel efficiency and decrease fuel consumption. The committee notes that initiatives such as re-engining aircraft, modifying in-flight profiles, and revising aircraft ground operations contribute to decreased fuel consumption and increased life-cycle savings.
>
> The committee is aware that winglet technology exists for aircraft to increase fuel efficiency, improve take-off performance, increase cruise altitudes, and increase payload and range capability. The committee notes that winglets are currently used on commercial aircraft and result in a five to seven percent increase in fuel efficiency. On September 16, 1981, the National

[1]NRC, 2007, *Improving the Efficiency of Engines for Large Nonfighter Aircraft*, Washington, D.C.: The National Academies Press.

> Aeronautics and Space Administration released the KC-135 Winglet Program Review on the incorporation of winglets for KC-135 aerial refueling aircraft. However, the Air Force concluded that the cost of adding winglets to the KC-135 did not provide sufficient payback in fuel savings or increased range to justify modification. Although the Air Force did conclude that modifying aircraft with winglets could increase fuel efficiency, the Air Force determined that re-engining the KC-135 aircraft produced a greater return on investment. The committee believes that incorporating winglets on military aircraft could increase fuel efficiency on certain platforms and that the Air Force should reexamine incorporating this technology onto its platforms.
>
> Therefore, the committee directs the Secretary of the Air Force to provide a report to the congressional defense committees by March 1, 2007, examining the feasibility of modifying Air Force aircraft with winglets. The report shall include a cost comparison analysis of the cost of winglet modification compared to the return on investment realized over time for each airlift, aerial refueling, and intelligence, surveillance, and reconnaissance aircraft in the Air Force inventory; the market price of aviation fuel at which incorporating winglets would be beneficial for each Air Force platform; all positive and negative impacts to aircraft maintenance and flight operations; and investment strategies the Air Force could implement with commercial partners to minimize Air Force capital investment and maximize investment return.

In response to a subsequent request from the Air Force, the NRC appointed the Committee on Assessment of Aircraft Winglets for Large Aircraft Fuel Efficiency to examine the feasibility of modifying Air Force aircraft with winglets. Since this study is a follow-on effort to the earlier study examining methods to improve fuel efficiency in large Air Force aircraft, appropriate members of the original study committee, including the chair and vice chair, agreed to participate in this study. They were joined by new members with the expertise to address the necessary technical areas. This report responds to the request of Congress as outlined above.

The chair thanks the members of the committee for generously taking time from their demanding schedules and working hard to complete this report in the short time allotted. The entire committee, in turn, thanks the many organizations and the guest speakers who provided excellent briefings and background information, and it thanks the NRC staff members who supported the study. Primary among them were Marta Vornbrock, Gregory Eyring, Jim Garcia, Michael Clarke, LaShawn Sidbury, and Detra Bodrick-Shorter.

Kenneth E. Eickmann, *Chair*
Committee on Assessment of Aircraft Winglets
for Large Aircraft Fuel Efficiency

Acknowledgments

This report has been reviewed in draft form by individuals chosen for their diverse perspectives and technical expertise, in accordance with procedures approved by the National Research Council's (NRC's) Report Review Committee. The purpose of this independent review is to provide candid and critical comments that will assist the institution in making its published report as sound as possible and to ensure that the report meets institutional standards for objectivity, evidence, and responsiveness to the study charge. The review comments and draft manuscript remain confidential to protect the integrity of the deliberative process. We wish to thank the following individuals for their review of this report:

William G. Agnew, NAE, General Motors Corporation (retired),
Kenneth C. Hall, Duke University,
Wesley L. Harris, NAE, Massachusetts Institute of Technology,
Frank T. Lynch, Independent Consultant, Yorba Linda, California,
Gregory S. Martin, GS Martin Consulting,
John P. Sullivan, Purdue University,
Charles F. Tiffany, NAE, The Boeing Company (retired), and
Henry T.Y. Yang, NAE, University of California, Santa Barbara.

Although the reviewers listed above have provided many constructive comments and suggestions, they were not asked to endorse the conclusions or recommendations, nor did they see the final draft of the report before its release. The review of this report was overseen by Alexander H. Flax, NAE. Appointed by the National Research Council, he was responsible for making certain that an independent examination of this report was carried out in accordance with institutional procedures and that all review comments were carefully considered. Responsibility for the final content of this report rests entirely with the authoring committee and the institution.

The committee also acknowledges and appreciates the contribution of the members of the Air Force Studies Board (AFSB) of the National Academies for their support of this study.

The AFSB, established in 1996 by the National Research Council at the request of the Air Force, brings to bear broad military, industrial, and academic scientific, engineering, and management expertise on Air Force technical challenges and other issues of importance to senior Air Force leaders. The board discusses potential studies of interest, develops and frames study tasks, ensures proper project planning, suggests potential committee members and reviewers for reports produced by fully independent ad hoc study committees, and convenes meetings to examine strategic issues. The board members listed on page vi were not asked to endorse the committee's conclusions or recommendations, nor did they review the final draft of this report before its release. Board members with appropriate expertise may be nominated to serve as formal members of study committees or to review reports.

Contents

Figures and Tables

FIGURES

TABLES

Acronyms

ACEE	aircraft energy efficiency
APB	Aviation Partners Boeing
API	Aviation Partners Incorporated
APU	auxiliary power unit
AWACS	Airborne Warning and Command System
BBJ	Boeing Business Jets
CFD	computational fluid dynamics
CG	center of gravity
DESC	Defense Energy Support Center
DOD	Department of Defense
DOE	Department of Energy
EGT	exhaust gas temperature
ESPC	energy savings performance contract
FAA	Federal Aviation Administration
ISR	intelligence, surveillance, and reconnaissance
JSTARS	Joint Surveillance Target Attack Radar System

L/D	lift-to-drag ratio
NACA	National Advisory Committee for Aeronautics
NASA	National Aeronautics and Space Administration
NG	Next-Generation
NPV	net present value
NRC	National Research Council
OEM	original equipment manufacturer
RN	Reynolds number
SOT	statement of task
TACAMO	take charge and move out
TOGW	takeoff gross weight
USAF	United States Air Force

Summary

Since the 1970s, when the price of aviation fuel began to spiral upward, airlines and aircraft manufacturers have explored many ways to reduce fuel consumption by improving the operating efficiency of their aircraft. Fuel economy concerns have been particularly keen for operators of commercial aircraft, which typically fly many hours per day in competitive markets, but they have been growing for military aircraft as well. The fuel consumed by the U.S. Air Force is in excess of 3 billion gallons per year, which is over 8 million gallons per day.[1] Aviation fuel accounts for much of this total, and the aircraft used by the Air Force for airlift, aerial refueling, and intelligence, surveillance, and reconnaissance (ISR)—which are the aircraft covered in this study—account for over half of all aviation fuel.[2]

One very visible action taken by commercial airframe manufacturers and operators to reduce fuel consumption is the modification of an aircraft's wingtip by installing, for example, near-vertical "winglets" to reduce aerodynamic drag. Experience shows that these tip devices reduce block fuel consumption (total fuel burn from engine start at the beginning of a flight to engine shutdown at the end of that flight) of the modified aircraft by

[1]Ron Sega, 2006, "Air Force energy strategy," Worldwide Energy Conference and Trade Show, Arlington, Va., April 19.

[2]Data provided by Defense Energy Support Center (DESC) on fuel usage by mission design series for FY05.

3-5 percent, depending on the trip length.[3] These wingtip modifications are offered as options to the original design of many newer commercial jetliners but are also available for retrofit to selected older aircraft. To date, however, only one military-unique aircraft (the C-17 transport) features winglets, though some studies have been conducted on the feasibility of retrofitting tip modification devices on other military aircraft.

In light of its growing concerns about rising fuel costs, the Air Force asked the National Research Council (NRC) to evaluate its aircraft inventory and to identify those aircraft that may be good candidates for winglet modifications. Specifically, the Air Force asked the NRC to perform the following four tasks:

1. Examine the feasibility of modifying Air Force airlift, aerial refueling, and ISR aircraft with winglets, to include a cost-effectiveness analysis of the feasible winglet modifications in net present value (NPV) terms.
2. Determine the market price of aviation fuel at which incorporating winglets would be beneficial for each platform.
3. Consider impacts to aircraft maintenance and flight operations (including ground operations).
4. Offer investment strategies the Air Force could implement with commercial partners to minimize Air Force capital investment and maximize investment return.

Although the statement of task above refers specifically to "winglets," the Committee on Assessment of Aircraft Winglets for Large Aircraft Fuel Efficiency chose to broaden the scope of its deliberations slightly by including a variety of possible modifications to the wingtip (e.g., wingtip span extensions). Thus, in this report, the term "winglet" denotes the traditional, nearly vertical wingtip design, while "wingtip modifications" is used to refer to the more general set of wingtip designs, including winglets and wingtip extensions, aimed at reducing aerodynamic drag.

These tasks call for a quantitative assessment of the costs and benefits of winglet modifications on a variety of platforms. In a comprehensive analysis, one would need to include the nonrecurring engineering costs of

---

[3]This range of 3-5 percent block fuel savings, derived from commercial experience, is lower than the 5-7 percent cited by the U.S. House of Representatives Armed Services Committee in Report 109-452, which may reflect fuel savings under cruise conditions.

wing analysis and wingtip design, the costs of materials, manpower, and out-of-service time to accomplish the modification, financial implications, training costs, potential impacts on maintenance docks and hangar space, costs associated with software and technical manual revisions, and any impacts on maintenance, operations, or mission accomplishment. Benefits to be considered would include not only improved fuel economy but also improved payload-range capability, improved takeoff performance, and less takeoff noise. In most cases, quantitative data on these costs and benefits were not known or not available. However, the committee did use preliminary NPV calculations to estimate payback periods for wingtip modification investments on various platforms by treating fuel costs, savings, and wing modification costs parametrically. These calculations supplemented the committee's expert judgment on which platforms appear to be the best candidates for wingtip modification.

Besides wingtip devices, there are other methods to reduce aircraft fuel consumption, but since they were not mentioned in the statement of task, the committee did not examine them in detail, nor did it examine the extent to which the Air Force might already be using some of these methods. Likewise, it did not make any formal recommendations concerning them. However, the committee suggests this is an area that should be considered as potentially providing significant fuel savings to the Air Force.

FINDINGS AND RECOMMENDATION

In this section, the committee presents two findings and a recommendation in response to the four tasks it was asked to perform.

Feasibility and Cost Effectiveness of Modifying Air Force Aircraft

Finding: The committee's analysis for a broad range of fuel prices and with the data available to it on potential improvements in block fuel savings, modification cost estimates, operational parameters for the aircraft, and so forth indicates that wingtip modifications offer significant potential for improved fuel economy in certain Air Force aircraft, particularly the KC-135R/T and the KC-10.

To assess the feasibility and cost-effectiveness of wingtip modifications, the committee began by investigating those aircraft in the Air Force inventory that burn the most fuel. In decreasing order of annual fuel burn (by

TABLE S-1 Potential for Wingtip Modifications to Benefit Air Force Aircraft

Aircraft	Priority/Potential Benefit
KC-10	High
KC-135R/T	High
C-5	Medium
C-17	Medium/low
C-130H/J	Low

fleet), they are the C-17, KC-135R/T, C-5, KC-10, and C-130H/J. Based on factors such as estimated fuel savings, cost of modification, operational flexibility, mission profiles, and remaining service life, the committee ranked these aircraft in order of their likely suitability for wingtip modifications, as shown in Table S-1.

KC-10

The KC-10 airframe is based on the commercial DC-10 airframe, and early commercial DC-10 flight tests validated a 2-3 percent improvement in fuel efficiency at cruise conditions with winglets as compared with the original wing design.[4] Not only was the DC-10 modified and tested with winglets, but its successor, the MD-11, was designed and certified with winglets. With the computational fluid dynamics (CFD) tools of today, moreover, a winglet or other wingtip modification designed for the KC-10 aircraft might well achieve greater fuel savings than were demonstrated on the DC-10 fitted with winglets some 25 years ago. In addition, recent winglet design experience using high Reynolds number (RN) wind tunnels could have applicability for winglet designs that may be more effective on the KC-10 and other government transport aircraft. As a result of all of this past work, the KC-10 fleet would require much less development time and effort to determine the effectiveness and suitability of various aerodynamic improvements.

KC-135R/T

The KC-135 airframe is closely related to the commercial Boeing 707. In the late 1970s and early 1980s, a joint National Aeronautics and Space

[4]A.B. Taylor, 1983, "DC-10 winglet flight evaluation summary report," NASA-CR-3748, December.

Administration (NASA)/Air Force program was conducted to evaluate the benefits that could be achieved from retrofitting winglets on the KC-135 aircraft. The wind tunnel test indicated that winglets would reduce KC-135 aircraft drag by 6-8 percent,[5] and flight tests with a KC-135 modified with winglets indicated substantial benefits. The study also indicated that the structural modifications required to install winglets on the KC-135s are a reasonable-size work package. Additional study would now be required to establish that the wings of these aging aircraft still meet the requirements of winglet installation.

C-5

Given that the C-5 is one of the largest contributors to the Air Force's fuel consumption and that its missions are long range, a study to quantify the potential gains and the effects of integrating winglets is warranted. Unfortunately, unlike the KC-10 and the KC-135, on whose derivative commercial aircraft there has been a comprehensive winglet development effort, a C-5 fleet retrofit program would add a measurable nonrecurring cost and require a longer time to recover the investment.

C-17

A number of design considerations led to the final winglet configuration on the C-17. One such consideration was that the wingspan was limited to that of the C-141 in order to maintain compatibility with facility infrastructure. The C-17 winglet was shown in wind tunnel testing to provide approximately 2.5 percent reduction in drag under cruise conditions. Also, flight testing showed no additional buffeting for takeoff or landing.[6] However, while these benefits are considered to be substantial, the C-17 winglet was developed in a low-RN wind tunnel. The low-RN environment can give misleading results with regard to drag, buffet, pitching moment, and loads because the much higher RN of the full-scale flight vehicle exhibits different flow phenomena. Also, the C-17 configuration was developed in the 1980s, before the full-scale wind tunnel at the National Transonic

---

[5]NASA, 1982, "KC-135 winglet program review," NASA Conference Publication 2111, January.

[6]Robb Gregg, Senior Manager for Aircraft Programs, Boeing Phantom Works, "Drag improvement: A study of the DC-10/MD-11/C-17 winglet programs," Presentation to the committee on December 13, 2006.

Facility became available and before modern Navier-Stokes CFD tools had been developed. With these new capabilities, a more accurate assessment of the current C-17 winglet design could be obtained. In addition, with these new tools and lessons learned from other winglet designs, it may be possible to improve the C-17 winglet design to reduce cruise drag another 1 percent or more.[7]

C-130H/J

Compared with the gains realized for commercial airline applications, the performance benefits provided by wingtip modifications on the C-130 would be less. For one thing, the C-130's wing is already very efficient because its aspect ratio of 10 is relatively high. Another reason for the lower gain in expected winglet efficiency is the C-130's unswept wing with its lower tip loading. In addition, since winglets are more effective for longer ranges and with the higher wingtip loading realized at higher altitudes, the potential benefit of winglets for the C-130 is limited.

Intelligence, Surveillance, and Reconnaissance Aircraft

While these aircraft are mentioned in the study's statement of task, the committee notes that they are not major fuel consumers, and their wings are already optimized for aerodynamic efficiency such that they would be expected to derive little benefit from wingtip modifications.

Other Air Force Aircraft

Finding: Most of the aircraft in the Air Force inventory that derive from commercial aircraft now operating with winglets already have winglets, or the decision has been made to install winglets. The remaining Air Force aircraft that are derivatives of commercial aircraft do not appear to be good candidates for wingtip modifications.

---

[7]Robb Gregg, Senior Manager for Aircraft Programs. Boeing Phantom Works, "Drag improvement: A study of the DC-10/MD-11/C-17 winglet programs," Presentation to the committee on December 13, 2006.

TABLE S-2 Winglet Status of Air Force Aircraft Derived from Commercial Airframes

Air Force Aircraft	Commercial Equivalent	Inventory	Winglets
C-9	Douglas DC-9-30	3	No
C-20B	Gulfstream GIII	5	Yes
C-20H	Gulfstream GIV, GIVSP	2	Yes
C-37	Gulfstream GV	9	Yes
C-21	Learjet 35A	59	No
C-40B	Boeing 737-700	4	Yes
C-40C	Boeing 737-700	3	Yes
VC-25	Boeing 747-200 (-300 wings)	2	No
E-4	Boeing 747-200	4	No
C-32	Boeing 757-200	4	Yes

SOURCE: Data provided by USAF.

The easiest decisions on whether to install winglets obviously involve aircraft in the Air Force inventory that derive from commercial aircraft now operating with winglets. In each case, the aircraft structure has already been studied and determined to be appropriate, the engineering design has been done, the modifications have been prototyped, tested, and certified, modification kits developed, flight manuals revised as required, and so on. However, the committee's review of all such Air Force aircraft revealed that most of them already have winglets, or the decision has been made to incorporate winglets, as shown in Table S-2.

All of these aircraft have winglets except for the C-9s, the C-21s, the VC-25s, and the E-4s. The three C-9s, derivatives of the DC-9, are scheduled to retire in FY11 and should not be considered for wingtip modifications. Also, past work on winglets for the DC-9, as discussed in Chapter 3, did not prove to be favorable. The C-21s, derivatives of the Learjet 35A, are small aircraft, so the entire fleet uses approximately 8 million gallons of fuel per year and would not be a priority for modification. Furthermore, they have tip tanks, and wingtip modifications would require the removal of these tanks, severely limiting the range of these aircraft even with a more efficient wing. Lastly, the VC-25s and the E-4s are derivatives of the Boeing 747-200, with the VC-25s having 747-300 wings. The 747-200 has not been produced since the late 1980s, so the commercial fleet is aging and retiring from service. As a result, the entire cost of winglets designed for 747-200/300 wings would have to be borne by the government. All of the Boeing 747s in the commercial world that have winglets are 747-400s, which have a structurally modified wing. The structural modification to allow installing the 747-400 wingtip on the VC-25s or the E-4s would be very expensive and impractical.

Preliminary Net Present Value Analysis

The committee followed up the qualitative analysis described above with a preliminary NPV analysis based on a simple spreadsheet model that considered a range of assumed modification costs and fuel savings for the most promising aircraft identified above. These preliminary NPV calculations confirm that wingtip modifications should be seriously considered for the KC-135R/T and KC-10 (see "Fuel Price Analysis," below). However, a detailed engineering and economic analysis would be required for each aircraft type before a final decision could be made to proceed with the installation of winglets or other aerodynamic modifications.

Recommendation: The Air Force should initiate an engineering analysis with the original equipment manufacturers (OEMs) to determine (1) the extent and cost of modifications needed for the KC-135R/T and the KC-10 to enable the installation of wingtip devices and (2) the fuel savings that could be achieved by this modification for each aircraft type. It should then perform an NPV analysis with these data to calculate the net savings. The Air Force should also analyze the C-5 and C-17 for potential wingtip modifications.

The OEMs have the detailed knowledge of wing designs and previous modifications that is necessary for carrying out these analyses. The results should be shared with the other Services operating similar aircraft.

Fuel Price Analysis

To illustrate the types of costs and benefits that might be realized through wingtip modifications (e.g., winglets) that would produce a reduction in fuel burn, the committee performed its own preliminary NPV analysis for the KC-135R/T and the KC-10. The analysis was used to determine whether wingtip modifications for selected aircraft would pay for themselves well before the aircraft are due to retire. Since it is not possible to know the modification costs and fuel savings without performing a detailed engineering analysis, these were treated as parameters in the model.

The range for modification costs was chosen from list prices and committee estimates. For fuel savings, the calculations were done for block fuel savings of 3 percent and 5 percent, consistent with commercial airline experience and the findings of this report. Results were calculated for the worst-case (highest modification cost and lowest fuel savings) and best-case (lowest modification cost and highest fuel savings) payback periods at a fuel cost of $2.50 per gallon. The committee assumed an annual fuel cost escalation rate of 3 percent and a discount rate of 3 percent.

In the KC-135R/T best case, net savings become positive 9 years after starting the modification program. All 417 aircraft in the inventory are modified. Total net savings to the Air Force are approximately $400 million (FY07 $). In the KC-135R/T worst case, net savings become positive 24 years after starting the modification program. Only 217 of the 417 aircraft in the inventory are modified (the others are not modified because they are expected to be retired from the inventory before reaching the end of their payback periods). Total net savings to the Air Force are approximately $36 million (FY07 $).

In the KC-10 best case, net savings become positive 8 years after starting the modification program. All 59 aircraft in the inventory are modified. Total net savings to the Air Force are approximately $221 million (FY07 $). In the KC-10 worst case, net savings become positive 23 years after starting the modification program. Only 53 of the 59 aircraft in the inventory are modified (the others are not modified because they are expected to be retired from the inventory before reaching the ends of their payback periods). Total net savings to the Air Force are approximately $12 million (FY07 $).

The price per gallon of fuel was also parameterized at $2.50, $5.00, $10.00, and $20.00 to account for the fully burdened cost of fuel. In constant dollars, when the cost of fuel is doubled, the payback period is cut in half. Total net savings to the Air Force rise significantly.

These numbers are illustrative only, and more accurate estimates of breakeven fuel prices would require engineering analysis to determine actual modification costs and the fuel savings potential for each aircraft.

Impacts on Aircraft Maintenance and Flight Operations

Commercial experience with aircraft that have installed winglets has shown that there have been no significant impacts on aircraft maintenance, flight operations, or ground operations (gate space, taxiways, hangars, etc.). Similarly, the Air Force has not experienced any significant impacts on aircraft maintenance or flight operations for aircraft it currently operates with winglets, and the committee does not expect any major problems with modifications to other aircraft under consideration.

Investment Strategies

Implementing the Modifications

Should the decision be to proceed with wingtip modification on the KC-10, the committee recommends the work be done while the aircraft are in normal scheduled overhaul. Since the KC-10 is maintained on contract with industry engineers who have intimate knowledge of commercial DC-10s, it is possible that wingtip modification could be added to the work specification with little or no added downtime or loss of operational availability.

Much of the same applies to the KC-135R/T aircraft fleet, except that unlike the KC-10, many of these aircraft are maintained by Air Force personnel in-house. The committee therefore believes that the wingtips could be retrofitted while the aircraft are undergoing their 5-year cycle of programmed depot maintenance. Rather than divert Air Force mechanics from other tasks, however, it might be wiser to partner with industry and have an experienced contract field team work with the Air Force mechanics to accomplish the modification. For the KC-135R/T undergoing programmed depot maintenance at contractor facilities, the Air Force should consider adding any proposed wingtip modifications to the existing overhaul contract. This would minimize training and allow returning the aircraft to the Air Force in the shortest possible time.

Financing Options

Wingtip modification programs and other fuel economy investments are examples of long-term investments that may require a significant initial investment that provides returns over time. Securing financing for such long-term investments is always a challenge given the current military acquisition practices and congressional appropriation processes. In a previous report on engine fuel economy in military aircraft,[8] the NRC discussed innovative financing mechanisms that might be pursued. The statement of task for that study included a request to "develop implementation strategies to include conventional, as well as innovative, acquisition, financing, and support concepts."[9] The committee believes that three of the mechanisms discussed in that report—specifically, creating a line item in the defense budget, implementing an "energy savings performance contract" strategy, and competing airframe maintenance contracts—could be applicable in implementing wingtip modifications. Those mechanisms are discussed in some detail in the earlier report.

CONCLUDING REMARKS

It is clear that aerodynamic improvements, including winglets, can make significant contributions to the efficiency of aircraft and should be considered for the military fleets discussed in this report. In each case, however, the appropriateness of such structural modifications must be determined fleet by fleet. These decisions are very complex and will depend on many factors, including the design of the aircraft structures, design margin within those structures, the condition of the structures, mission profiles, utilization rates, fuel consumption rates, fuel prices, and the remaining life of the aircraft. The Air Force should support the analysis required and make the appropriate modifications as quickly as possible. There are also other ways to reduce fuel consumption, many of which have already been adopted by the commercial airlines. The committee believes it is important for these other strategies to be considered, and while they were not the focus of this study and the extent to which the Air Force may already be using some of these strategies was not examined, examples are provided in Appendix B for the reader's benefit.

---

[8]NRC, 2007, *Improving the Efficiency of Engines for Large Nonfighter Aircraft*, Washington, D.C.: The National Academies Press.

[9]Ibid.

1

Background and Overview

INTRODUCTION

As discussed in the Summary, airlines and aircraft manufacturers have been particularly keen on reducing fuel consumption given increasing fuel prices and today's competitive markets. However, the fuel economy of military aircraft has become an increasing concern as well. The fuel consumed by the U.S. Air Force is in excess of 3 billion gallons per year, which is over 8 million gallons per day.[1] Aviation fuel accounts for much of this total, and the aircraft used by the Air Force for airlift, aerial refueling, and intelligence, surveillance, and reconnaissance (ISR)—which are the aircraft covered in this study—account for over half of all aviation fuel.[2] The stated Air Force policy is now to "make energy a consideration in all Air Force actions" and to "promote a culture in which airmen conserve energy."[3] More generally, reduced energy consumption and reduced dependence on foreign oil have become strategic goals of the U.S. Department of Defense (DOD).[4]

Broadly speaking, the fuel economy of an aircraft can be thought of as having three components: the efficiency of the engines, the aerodynamic

[1]Ron Sega, 2006, "Air Force energy strategy," Worldwide Energy Conference and Trade Show, Arlington, Va., April 19.

[2]Data provided by Defense Energy Support Center (DESC) on fuel usage by mission design series for FY05.

[3]Ibid.

[4]Terry Pudas, 2006, "A strategic approach to energy," *Defense Technology International* May/June:42.

FIGURE 1-1 A common wingtip modification is the "winglet." SOURCE: Aviation Partners Boeing (APB).

performance, and the weight efficiency. In a recent report, the National Research Council (NRC) examined the potential for improving engine performance in military aircraft and briefly discussed various aerodynamic improvements.[5] This report examines potential aerodynamic improvements in large military tanker and transport aircraft in greater detail, in particular the potential for the modification of the wingtips to reduce aerodynamic drag. An example of such a wingtip modification is the "winglet" now seen on many commercial jet aircraft and some military aircraft, shown in Figure 1-1; however, many other aerodynamic improvements are possible.

The concept of winglets was originally developed in the late 1800s by British aerodynamicist F.W. Lancaster, who patented the idea that a verti-

[5]NRC, 2007, *Improving the Efficiency of Engines for Large Nonfighter Aircraft*, Washington, D.C.: The National Academies Press.

cal surface at the wingtip would reduce drag.[6] The idea was refined in the late 1970s at the National Aeronautics and Space Administration (NASA) Langley Research Center by Richard Whitcomb, who designed a winglet using advanced airfoil concepts integrated into a swept, tapered planform that would interact with the wingtip airflow and circulation to reduce drag. Dr. Whitcomb proved the efficacy of winglets in wind tunnel and computer studies.[7]

The first commercial aircraft to use winglets were corporate-size Learjets in 1977, and the first big jetliner to feature winglets was the Boeing 747-400, followed by the MD-11.[8] Winglets and wingtip modifications are now standard equipment on many business jets and jetliners (e.g., Airbus A320/330/340/380; Boeing 747-400). In addition, winglet options are now offered on Boeing 737 aircraft. Winglets are also original equipment on the C-17 military transport. Winglet retrofit kits and services are available for the modification of older aircraft.[9]

Besides improved fuel economy—which tests suggest may be as high as 5 percent (this may be traded off to obtain increased range)—aircraft manufacturers and winglet retrofit companies have reported that winglets also offer higher operating altitudes, improved aircraft roll rates, shorter time-to-climb rates, lower takeoff speeds, and less takeoff noise. In the commercial world, winglets have not only reduced fuel costs but have also increased operational flexibility by, for example, bringing new international destinations within range and increasing payload capability at airports at high altitudes or with shorter runways.

The payback time for wingtip modification investments in large military tankers and transport aircraft is likely to be longer than the time for the corresponding commercial aircraft, since on average these military aircraft fly many fewer hours per year than do commercial jetliners. However, in combination with fuel savings, the ancillary operational flexibility offered by winglets may make a winglet retrofit a good idea for certain types of military aircraft. This is the issue examined in this report.

---

[6]Joseph R. Chambers, 2003, *Concept to Reality: Contributions of the Langley Research Center to U.S. Civil Aircraft of the 1990s*, NASA SP-2003-4529. Available online at http://oea.larc.nasa.gov/PAIS/Concept2Reality. Last accessed on February 26, 2007.

[7]Ibid.

[8]Ibid.

[9]Aviation Partners Boeing.

STATEMENT OF TASK

As noted in the preface, this report follows up on an earlier NRC study requested by the U.S. Air Force dealing with the re-engining of military aircraft. The following four tasks are addressed in this report:

1. Examine the feasibility of modifying Air Force airlift, aerial refueling, and intelligence, surveillance, and reconnaissance (ISR) aircraft with winglets, to include a cost-effectiveness analysis of the feasible winglet modifications in net present value (NPV) terms.
2. Determine the market price of aviation fuel at which incorporating winglets would be beneficial for each platform.
3. Consider impacts to aircraft maintenance and flight operations (including ground operations).
4. Offer investment strategies the Air Force could implement with commercial partners to minimize Air Force capital investment and maximize investment return.

SCOPE AND COMMITTEE APPROACH

Although the statement of task (SOT) specifically uses the term "winglet," which typically refers to a nearly vertical surface located at the wingtip, the committee chose to broaden the scope of its deliberations slightly to include a variety of possible modifications to the wingtip (e.g., wingtip span extensions) that can have a similar impact on fuel economy and aerodynamic performance. Thus, in this report, winglet denotes the traditional, nearly vertical wingtip design, while "wingtip modifications" will be used to refer to the more general set of wingtip designs, including winglets and wingtip extensions, aimed at reducing aerodynamic drag. In addition, given the SOT's emphasis on fuel economy, the committee also considered a variety of possible aerodynamic modifications and operational changes to the aircraft (e.g., improved pressure seals, improved control systems) that would be expected to be relatively simple and inexpensive to implement and that, taken together, might provide fuel economy benefits comparable to those provided by wingtip modifications. Since they were outside its charter, the committee did not examine these other methods in detail, nor did it examine the extent to which the Air Force might already be using some of these methods. Likewise, it did not make any formal recommendations concerning them. However, the committee suggests this is an area that should be considered as potentially providing significant fuel savings to the Air Force.

The committee also recognized that some of the other reported benefits of wingtip modifications, such as increased range and endurance, ability to utilize shorter runways, increased payload, and decreased time to climb, might be particularly valuable for certain Air Force missions, and that wingtip modifications might therefore be justified for reasons other than fuel cost savings.[10] While it was not possible to quantify these benefits exactly, the committee sought to consider them qualitatively in its assessment.

In tackling Task 1, the committee first generated a list of all Air Force aircraft that would be candidates for retrofit wingtip modifications. The committee assessed the missions and typical flight profiles of those that do not currently have wingtip modifications to identify the most promising subset of aircraft to subject to a more detailed analysis. Based on the testimony of representatives of aircraft manufacturers and on information provided by the Air Force, the committee sought to determine qualitatively the cost—including the cost of engineering analysis, structural modification to the wing, and so forth—of retrofitting each system compared to the fuel savings predicted to accrue. For aircraft that already have wingtip modifications, the committee assessed whether further aerodynamic improvements for even more fuel efficiency were warranted.

Task 2 seeks to determine a price for aviation fuel at which the cost of wingtip modification retrofits is justified by fuel cost savings alone. For the most promising subset of aircraft identified in Task 1, the committee estimated the cost of wingtip modification retrofit based on the estimated cost of retrofitting comparable commercial aircraft. By also estimating the potential fuel savings and the number of these aircraft, the committee performed preliminary NPV calculations to calculate whether wingtip modifications for selected military aircraft would pay for themselves well before the aircraft are due to be retired. Recognizing that the cost of fuel delivered to the location where it is used may be many times higher for military aircraft than for commercial aircraft,[11] the committee treated fuel cost as a parameter that could be varied over a large range.

---

[10]Some of these benefits, such as increased payload and range, must be traded off for fuel savings.

[11]AFSAB (Air Force Scientific Advisory Board), 2006, *Technology Options for Improved Air Vehicle Fuel Efficiency: Executive Summary and Annotated Brief*, SAB-TR-06-04, May.

As required by Task 3, the committee considered the impact of wingtip modifications on maintenance (depot and field) and flight operations (including hangars, runways, taxiways, and mission requirements), basing its analysis on experience with comparable commercial aircraft.

For those Air Force aircraft that the committee judged were the most promising candidates for wingtip modifications, the committee suggests investment strategies, as called for by Task 4.

STRUCTURE OF THIS REPORT

Chapter 2 discusses how wingtip modifications work, including how they affect aerodynamic performance. It identifies the various benefits and potential negative impacts of wingtip modifications. Chapter 3 summarizes the commercial and military experience with wingtip modifications, as well as lessons drawn from past studies and experience. In Chapter 4, the committee identifies the Air Force aircraft it found to be the best candidates for wingtip modifications. This is followed by a qualitative analysis of the relative costs and benefits of retrofitting wingtip modifications on these aircraft, as well as a discussion of appropriate strategies the Air Force should use to maximize its fuel economy investments. Additional methods that might be considered by the Air Force to improve fuel economy, such as other aerodynamic changes, improving maintenance and operations, and reducing unnecessary weight, are discussed in Appendix B.

2

Wingtip Modifications

HISTORY OF WINGTIP DEVICES

Within a few years of the first heavier-than-air flight, the idea of beneficial wingtip devices was introduced. Lanchester patented the concept of a wing end plate in 1897 and suggested that it would reduce wing drag at low speeds. Theoretical studies of end plates by Munk in 1921[1] were followed by studies of von Karman and Burgers[2] and Mangler[3] in the 1930s, a patent on nonplanar wings was granted to Cone in 1962,[4] and a paper on the topic was published by Lundry and Lissaman in 1968.[5] This work was paralleled by many experimental studies (see, for example, National Advisory Committee for Aeronautics (NACA) work from 1928[6] to 1950[7]), most of which

---

[1]M.M. Munk, 1921, "The minimum induced drag of aerofoils," NACA Report 121.

[2]T. von Karman and J.M. Burgers, "General aerodynamic theory—perfect fluids," In *Aerodynamic Theory,* W.F. Durand, ed., Berlin/Vienna: Julius Springer-Verlag, 1934-1936, and New York: Dover Publications, 1963, Div. E, Vol. II, pp. 216-221.

[3]W. Mangler, 1938, "The lift distribution of wings with end plates," NACA TM 856; transl. by J. Vanier from "Die Auftriebsverteilung am Tragflügel mit Endscheiben," *Luftfahrtforschung* 14:64-569.

[4]C.D. Cone, Minimum Induced Drag Airfoil Body, U.S. Patent 3,270,988, September 1966.

[5]J.L. Lundry and P.B.S. Llssaman, 1968, "A numerical solution for the minimum induced drag of nonplanar wings," *Journal of Aircraft* 5(1).

[6]Paul E. Hemke, 1928, "Drag of wings with end plates," NACA TR-267.

[7]John M. Riebe and James M. Watson, 1950, "The effect of end plates on swept wings at low speed," NACA TN-2229.

did not attain the potential savings suggested by the theory. This was partly due to simplistic design, which often included low-aspect-ratio, untwisted, flat-plate airfoils. Recognition of the importance of winglet location, twist, and aspect ratio was clear in the patent of Vogt in 1951[8] and in a variety of other nonplanar wingtip geometries studied and patented by Cone.[9] In the early 1970s, Whitcomb[10] of the National Aeronautics and Space Administration (NASA) defined and tested high-aspect-ratio, carefully designed nonplanar wingtips, termed "winglets," which were soon to appear on numerous aircraft, including Rutan's VariEze in 1975 and the Learjet 28/29 in 1977. The winglet of the Boeing 747-400 has a much lower dihedral angle than the Whitcomb winglet, and since that time, numerous vertical, canted, and horizontal wingtip extensions have been put into commercial and military service, as shown in Figure 2-1.

INTRODUCTION TO WINGTIP AERODYNAMICS

Much of the drag of an aircraft is related to the lift generated by its wing. To create this lift, the wing pushes downward on the air it encounters and leaves behind a wake with a complex field of velocities. This air behind the wing moves downward then outward, while the air outboard of the wing tips moves upward, then inward, forming two large vortices, as shown in Figure 2-2.

The energy required to create this wake is reflected in the airplane's "induced" or "vortex" drag. For most aircraft, induced drag constitutes a large fraction, typically 40 percent, of cruise drag. During takeoff, induced drag is even more significant, typically accounting for 80-90 percent of the aircraft's climb drag. And while takeoff constitutes only a short portion of the flight, changes in aircraft performance at these conditions influence the overall design and so have an indirect, but powerful, effect on the aircraft's cruise performance. Consequently, concepts that reduce induced drag can have significant effects on fuel consumption.[11]

---

[8]Richard Vogt, Twisted Wing Tip Fin for Airplanes, U.S. Patent 2,576,981, December 1951.

[9]C.D. Cone, Minimum Induced Drag Airfoil Body, U.S. Patent 3,270,988, September 1966.

[10]Richard T. Whitcomb, 1976, "A design approach and selected wind-tunnel results at high subsonic speeds for wing-tip mounted winglets," NASA TN D-8260.

[11]Ilan Kroo, 2005, "Nonplanar wing concepts for increased aircraft efficiency," VKI Lecture Series on Innovative Configurations and Advanced Concepts for Future Civil Aircraft, June 6-10.

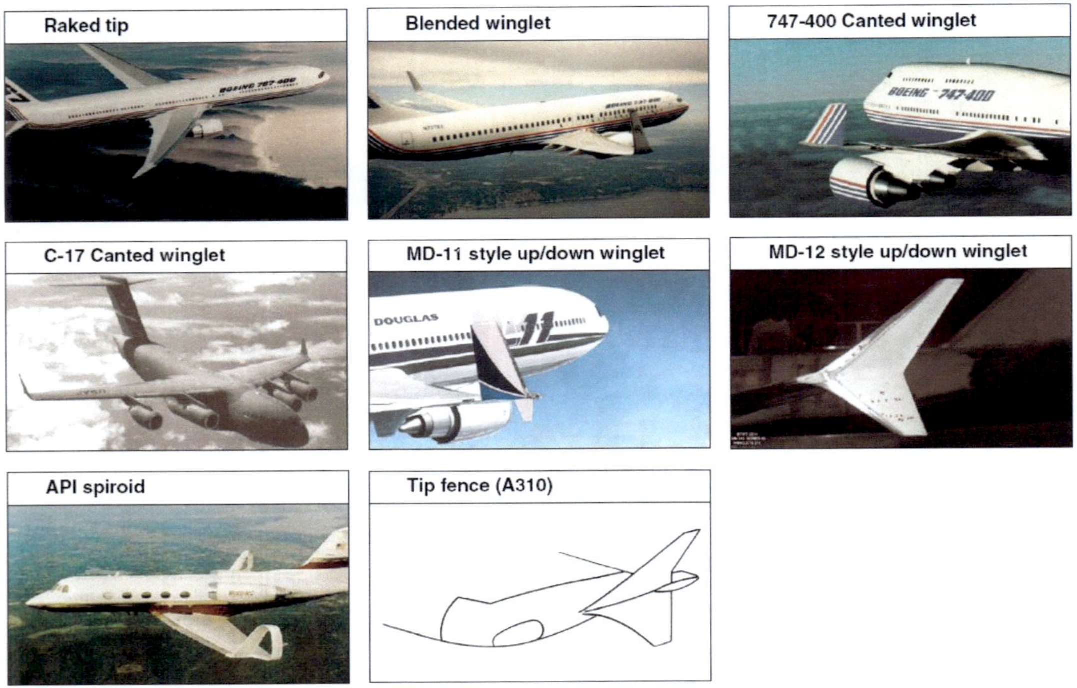

FIGURE 2-1 Wingtip modifications with a variety of geometries have been tested and deployed on both commercial and military aircraft. SOURCE: Courtesy of Doug McLean, "Wingtip devices: What they do and how they do it," Presentation at the Boeing Performance and Flight Operations Engineering Conference, September 2005.

FIGURE 2-2 The vortex wake behind lifting wings descending through a thin cloud layer. SOURCE: Airliners.net. Photo courtesy of S.C. Morris.

Note that the wake flow pattern illustrated in Figure 2-2 is a gross feature of the wing lift generation, not a localized phenomenon associated with wingtip geometry, so that reduction of the induced drag requires more than a small "device" at the tip. The basic method by which the vortex drag may be reduced is to increase the horizontal or vertical extent of the wing: By increasing the wing dimensions, a larger mass of air can be affected by a smaller amount to produce a given lift, and this leads to less energy in the wake and lower induced drag. So, perhaps the simplest means to reduce induced drag is to increase wingspan through horizontal wingtip extensions. However, in some cases this modification may not be appropriate because of explicit geometric constraints such as hangar width; in others it may not be desirable because of the increased structural weight of the wing, which must be designed to carry greater bending loads. On the other hand, adding vertical wing extensions creates many of the same effects as increasing the wing span (although one must add a bit more than twice the length of wing vertically to achieve the same savings as a horizontal extension). Vertical wing extensions (e.g., winglets) increase the effective span of the wing, lowering induced drag but increasing wing bending moments. They impose different and sometimes more acceptable challenges than horizontal wingtip extensions.

DESIGN OF WINGTIP DEVICES

Winglets are a visible sign of an improvement that is often perceived as high technology, and this apparently appeals to a segment of the commercial customer community. But from an aerodynamicist's point of view, the motivation behind most wingtip devices is to reduce induced drag. Beyond that, as Whitcomb showed, the designer's job is to configure the device so as to minimize the offsetting penalties, resulting in a net performance improvement. There are also aerodynamic and structural aspects that must be considered in the design of the wingtip device. The performance improvement for any particular wingtip device can be measured relative to the performance of the same airplane with no tip device.

Aerodynamic factors potentially offsetting these induced drag savings include an increase in the profile drag due to increased wetted area and junction flows, high sectional loadings, and so on and an increase in the trim drag resulting from increased outboard loading. The amount of trim drag increase is dependent on the specific aircraft and the ability to control the cruise center-of-gravity location (e.g., via fuel management). Increased outboard loading also increases the deflection of the wing at cruise, reducing the drag benefit relative to using a tip device on a theoretical rigid wing. Thus, the benefit associated with the tip device will depend on the specific aircraft and the structural margins of the wing. Finally, the wingtip device adds weight that comes not only from the device itself and its attachment fitting but also from any structural modifications to the existing wing to allow it to handle the additional static loads and to meet flutter and fatigue requirements.

Optimal Wingspan

As stated earlier, induced drag can usually be reduced by simply increasing wingspan, with a resulting reduction in total fuel consumption. Why, then, do aircraft have the limited spans they do if larger spans almost always reduce drag? There are two principal reasons for this:

- Aircraft are often span-constrained due to infrastructure and operational considerations such as hangar, gate, or taxiway dimensions.

For instance, the A380 was limited to a 262.5-ft span to be compatible with large airport infrastructures. Naval aircraft are often span-constrained by aircraft carrier elevator dimensions and deck limitations.

- Larger spans generally entail larger structural loads on the wings and therefore increased material and manufacturing costs. Eventually, the increased structural weight offsets the drag advantage of larger spans, but simple scaling laws suggest that this does not occur until the wings weigh about one-third as much as the total airplane. Nonetheless, the increased weight and cost of larger span wings leads to diminishing returns as span is increased. This, combined with the geometric issues noted above, determines the optimal span.

Many aircraft in the Air Force inventory were designed at a time when fuel costs were far lower than they are today—especially when the fully burdened cost of delivered fuel is considered. However, as fuel costs increase, the optimal span increases, since the ratio of fuel cost to manufacturing costs becomes larger. This means that if these same aircraft were being designed today, their spans would likely be larger than those of the aircraft in the current fleet.

To improve fuel economy, several options are possible. One could, for example, buy new aircraft designed for current and future fuel costs; redesign the wings of the most widely used aircraft and re-wing the existing airframes; or modify just a portion of the existing wings (by installing a retrofit device) to achieve a portion of the potential fuel savings. Retrofitting existing wings may be the lowest cost option in the near term. This option is especially attractive for aircraft having substantial structural margins.

Wing Retrofits

Several approaches to wing retrofits, which increase the effective aerodynamic span, are possible. The addition of winglets is perhaps the most obvious approach—obvious because of the recent success of winglet retrofits for the Boeing 737s and 757s and because the effective span may be increased without changes to the geometric span. Simple wingtip span extensions are also viable alternatives for reducing fuel consumption. Rather than adding winglets with a height of 10 ft, one could add 5-ft horizontal span extensions to each wingtip and achieve similar drag savings. Span extensions have been added to many commercial aircraft such as the DC-9, the DC-10, and the Boeing 767. They are less obvious than winglets but can also reduce fuel consumption and, depending on the details of the original design, may be more effective. Some aircraft growth versions have included both tip extensions and root plugs (DC-9 Series 50 to MD-80).[12] This approach involves more substantial modification of the wing but can produce greater fuel savings than simple tip modifications, adding wing area and permitting higher root bending loads than would be possible with tip changes alone.

Whether a specific existing wing is best modified by adding winglets or wingtip span extensions depends on many factors. If an aircraft is span constrained, a well-designed winglet can provide a significant reduction in drag. However, if an aircraft is not span constrained, whether to use winglets or tip extensions is less clear. Both winglets and tip extensions add bending loads, subsequently increasing the wing weight. In one study allowing for identical increases in root bending moments, winglets produced better results than tip extensions.[13] However, in another study in which integrated bending moments were constrained, winglets and tip extension produced the same results.[14] Both of these studies employed highly simplified models of the wing structure. In practice, the existing structure and load distribution must be considered. If, for example, substantial structural margins are available on the outer portion of a wing (e.g., due to minimum gauge constraints) but little is available at the root, a winglet might be added more easily than a span extension.

The geometry of the best wing extension or winglet retrofit also depends on other critical structural constraints. If flutter is critical, the reduced torsional frequencies created by winglets may lead to the choice of a smaller horizontal extension. Similarly, if large sideslips at high dynamic pressure are required for military operation, winglet loads could exceed loads of conventional span extensions. These various constraints make it difficult to generalize about winglets versus tip extensions. Also, stability and control changes can often be accommodated with either modification,

---

[12]The terms "root plug" and "root insert" refer to a modification to a wing in which span is added at the inboard end of the wing, adjacent to the fuselage. This is similar to a tip extension, which is added at the wingtip. For example, the MD-80 uses both a root insert and a tip extension.

[13]H.H. Heyson, G.D. Riebe, and C.L. Fulton, 1977, "Theoretical parametric study of the relative advantages of winglets and wing-tip extensions," NASA Technical Paper 1020.

[14]R.T. Jones and T.A. Lasinski, 1980, "Effects of winglets on the induced drag of ideal wing shapes," NASA Technical Memorandum 81230.

but as with structural considerations, they must be treated in detail on a case-by-case basis.

BENEFITS OF WINGTIP MODIFICATIONS

A net aerodynamic performance improvement made possible by wingtip modifications is satisfying to an engineer, but for an airplane manufacturer or operator the objective is to realize the kind of bottom-line benefits that translate into real savings as measured by cost, noise, engine exhaust emissions, operational flexibility, etc. The potential bottom-line benefits of wingtip devices are reduced fuel burn, increased capability, and improved performance, described below in order of importance.

Reduced Fuel Burn

By reducing drag, wingtip devices help the aircraft operate more efficiently and, in turn, reduce fuel burn. The fuel savings benefits of wingtip modifications depend on the mission flight profile, particularly the range and time spent at cruise speed. Commercial experience with winglet retrofits on the Boeing 737-300/700/800 indicate a 1.5 percent block fuel savings for trips of 250 nautical miles (nmi), increasing to 4 percent for trips of 2,000 nmi.[15] For the Boeing 757-200 and 767-300, block fuel savings were 2 percent for 500 nmi trips and 6 percent for 6,000 nmi. On an annual basis, winglets were projected to result in savings to commercial operators of up to 130,000 gallons of fuel per aircraft on the 737-800 and up to 300,000 gallons per aircraft on the 757-200.[16] Reduced fuel consumption translates directly into a reduction in operating cost.

Increased Payload-Range Capability

If less fuel is required to accomplish a particular mission at a specific takeoff weight, then that credit can be realized in more than one way. For example, the aircraft can carry more weight (more payload) the same distance or it can carry the same payload farther (greater range). Figure 2-3 shows the increase in payload-range capability made possible by winglets on

[15]Jay Inman, Vice President of Programs, Aviation Partners Boeing, "Blended winglets," Presentation to the Committee on Analysis of Air Force Engine Efficiency Improvement Options for Large Non-fighter Aircraft on June 14, 2006.

[16]Ibid.

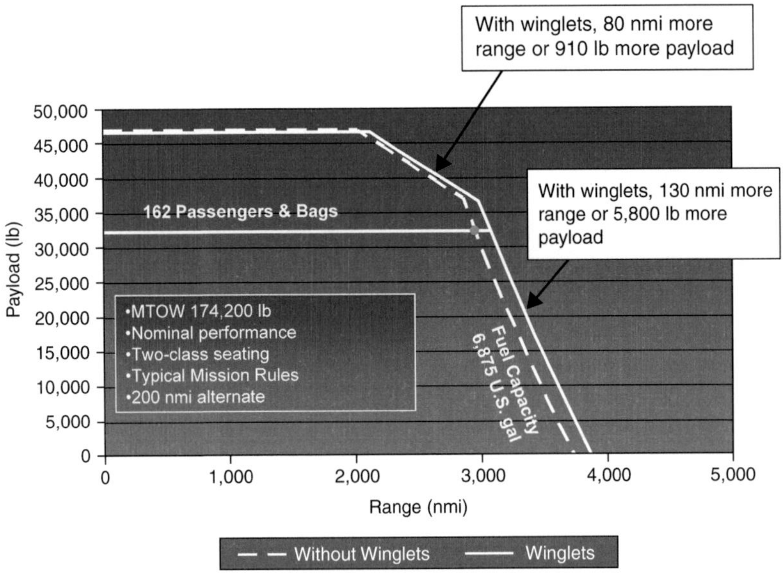

FIGURE 2-3 Winglets increase payload-range capability of the Boeing 737-800. SOURCE: Aviation Partners Boeing, Presentation to the Committee on Analysis of Air Force Engine Efficiency Improvement Options for Large Non-fighter Aircraft on June 14, 2006.

one commercial aircraft, the Boeing 737-800. The benefits begin to become apparent for ranges beyond 2,000 nmi. Between the 2,000 and 3,000 nmi range, winglets enable 80 nmi more range or 910 lb more payload. Beyond the 3,000 nmi range, winglets allow for 130 nmi more range or 5,800 lb more payload.[17] In the commercial world, this capability translates into operational flexibility—for example, it offers a greater choice of aircraft along certain routes or the opening up of new routes and destinations that were not previously within range.

The increased payload-range capability is valued in military aircraft applications just as it is in commercial aircraft applications. Carrying more payload to the same distance could mean fewer sorties to accomplish a specific goal, or it could allow servicing more customers with the same number of operational aircraft.

[17]Ibid.

Improved Takeoff Performance

The reduced drag associated with wingtip modifications reduces the thrust levels required for takeoff (reducing community noise at the same time) and enables faster second-segment climb. This increased climb rate allows the use of airports having shorter runways and allows for operations from airports located at higher altitudes and in hotter climates. Alternatively, these advantages may be traded for carrying higher payloads or a combination of both.

Critical performance constraints for military aircraft can be dictated by either airfield constraints or a combat situation. For example, at an airfield in hostile territory, a steep climb out may be desired to reduce the time an aircraft is vulnerable to surface-to-air threat systems around the airfield. Another example would be takeoff and landing constraints at a commercial airport where military tankers, airlift, or ISR platforms may also have to operate.

CHALLENGES ASSOCIATED WITH WINGTIP MODIFICATIONS

The potential benefits of wingtip modifications do not come without a price. Offsetting factors include the cost of the modification, added weight, added span and height, and potential interference with other wing equipment. These offsetting considerations are discussed below.

Cost

The costs of a wingtip modification retrofit include the nonrecurring costs for engineering, for modification of the wing itself, and for tip device design, manufacturing, and installation. To determine if a wingtip modification is cost-effective, the extent and cost of the nonrecurring engineering and of modifying the existing wing must be calculated. The wing modification costs depend on specific wing characteristics, including structural margins and loadings, as well the strength remaining in light of structural fatigue and corrosion. The wing modifications required to accommodate a tip device could be extensive.

Currently, a winglet retrofit kit for a suitable narrow-body commercial jetliner like the Boeing 737 costs from $500,000 to $1 million per aircraft. For a wide-body like the Boeing 767, the costs are between $1 million and

$1.5 million. For a jumbo-sized aircraft like the Boeing 747, the costs would probably be higher.[18]

A military aircraft having a close commercial analogue that has been evaluated or fitted with tip devices could have substantially lower nonrecurring engineering costs because of this existing knowledge. For example, the C-32 is based on the Boeing 757-200, which has already been modified with winglets; therefore, that experience can inform the decisions regarding the C-32.[19] Similarly, in the 1980s the suitability of winglets on the KC-135 was studied.[20] This previous work could help to inform a winglet retrofit decision today. However, the KC-135s are now more than 20 years older, and the current condition of their wings would need to be evaluated.

Winglets may have a smaller nonrecurring statement of work than other means of achieving similar improvements. For example, a re-engine program can also improve fuel burn, operational flexibility, and takeoff performance. If the magnitude of the needed improvements is similar, the winglet solution would almost certainly be less costly.

Added Weight

There are two components of added weight: (1) any modifications to the wing that might add weight (e.g., stiffening of the wing to satisfy static and dynamic requirements) and (2) the weight of the winglets themselves. As examples, commercial designs have yielded total modification weights (winglet plus wing modification) of 340 lb for the 737-700 and 1,358 lb for the 757-200ER.[21]

Added Span and Height

The height of a winglet varies but can be as great as 10-20 ft. A winglet can also increase the wingspan by several feet. These dimensions impact airfield operations such as parking, taxiing, and maneuvering the aircraft on

[18]Costs based on list prices and committee estimates.

[19]The C-32 was recommended for winglet modification in the DOD Appropriations Bill for 2007 and signed into law on September 29, 2006.

[20]NASA, 1982, "KC-135 winglet program review," NASA Conference Publication 2111, January.

[21]Jay Inman, Vice President of Programs, Aviation Partners Boeing, "Blended winglets," Presentation to the Committee on Analysis of Air Force Engine Efficiency Improvement Options for Large Non-fighter Aircraft on June 14, 2006.

the ground. If space is critical, a few additional feet of span per aircraft could limit the number that can be on an airfield at any given time, also known as "maximum on ground." This could constrain throughput for cargo and tanker aircraft, in particular. Winglet height could be an issue if there are obstacles that the winglet would hit when parking or taxiing, damaging both the winglet and obstacle.

However, winglets may be more compatible with existing infrastructure than, say, wingtip extensions. For the same aerodynamic improvement, winglets typically add less span to the airplane than a wingtip extension and might enable the continued use of existing ramp space, gates, hangars, etc.

Interference with Other Wing Equipment

Wingtip modifications might also impact other wing requirements. For example, a winglet might interfere with antennas or sensor equipment on military airplanes. Wingtip modifications might also impact airplane lighting solutions, anti-icing system requirements, and lightning strike dissipation solutions. Winglets can be efficient ice collectors and raise ice protection issues. Such problems should be thoroughly assessed before committing to any wingtip modification solution. Also, wingtip modifications may alter the effectiveness of high lift or control devices by changing their aerodynamic loading either favorably or adversely. Wings with outboard lateral control devices (ailerons, spoilers, and the like) may be particularly susceptible to changes resulting from the addition of a wingtip device such as a winglet or a wingtip extension.

GENERAL OBSERVATIONS

In summary, there are many questions that have to be answered and trade-offs that have to be evaluated in determining whether or not to invest in wingtip modifications. Do wingtip devices require that the wing be strengthened in order, for example, to deal with added moments that might be introduced by wingtip modifications? What is the work package that needs to be developed to assess the extent of the modification, the cost of the modification, and the time an aircraft is out of service? What is the remaining life of the aircraft over which the costs will be amortized? All of these factors will determine the overall costs, which can then be compared with the overall benefits in order to decide whether to go forward. One cannot simply say that because wingtip modifications save fuel on commercial aircraft, the Air Force should embark on putting wingtip modifications on its mobility aircraft. Investigating the viability/efficacy of such modifications is of value, and a lot can be learned from the extensive work that has already been done on commercial aircraft. But one should not assume a priori that such an investigation will result in a decision to proceed. Both engineering and operational analyses must be done to inform an investment decision.

3

Previous Analyses and Experience with Wingtip Modifications on Existing Aircraft

This chapter reviews the results of previous studies and deployment of wingtip devices on existing commercial and military aircraft. On the commercial side, the experience and decision processes of aircraft manufacturers as well as two operators (airlines) are described. On the military side, one transport aircraft (the C-17) already has winglets as original equipment, and others (e.g., the KC-135 tanker) have been evaluated for this modification. In some cases, military airplanes are closely related to commercial analogues (e.g., the C-32 is based on the Boeing 757-200) and can benefit from wingtip modification studies that have been conducted for the commercial aircraft. In other cases, the aircraft are military-unique (e.g., the C-5), and the evaluation of their suitability for wingtip modifications would have to start from the beginning. Military-unique aircraft for which no previous wingtip modification studies are available are discussed in the next chapter. This chapter concludes with a summary of the lessons that can be learned from past experience with wingtip devices that could help with deciding on military wingtip device retrofits.

EARLY RESEARCH PROGRAMS

NASA-Led Research: ACEE

The Aircraft Energy Efficiency (ACEE) program was initiated by the Department of Energy (DOE) and the National Aeronautics and Space

Administration (NASA) in 1975 as a 10-year effort to advance aircraft performance and increase fuel economy by 40 to 50 percent per unit distance. The technological opportunities for doing this included more fuel-efficient engines, lighter weight structures, and better aerodynamic designs.[1]

The winglet concept was evaluated and tested extensively by NASA in its 8-Foot Transonic Pressure Tunnel from 1974 to 1976. In July 1976, NASA published a general design approach that summarized the aerodynamic technology involved in winglet design. The tunnel tests indicated that, for typical subsonic transport aircraft configurations, induced drag could be reduced by about 20 percent and the aircraft lift-to-drag ratio (L/D) could be increased by about 9 percent. This improvement in L/D is more than twice as great as that achieved by the comparable wingtip extension producing the same wing-root bending moment.[2,3] In coordination with the ACEE program, Boeing, McDonnell Douglas, and Lockheed studied the impact of winglets on near-term derivative aircraft. The results of these efforts and additional efforts by the airframe manufacturers are described below.

COMMERCIAL EXPERIENCE

There are a number of very successful applications of winglets and wingtip extensions in the world's commercial airplane fleet. These programs have been successful for a number of reasons, most notably because they have enhanced the economic value of the subject commercial airplanes. These wingtip device strategies have been employed both on new design aircraft and as postproduction retrofits on existing aircraft. The following is a summary of the strategies employed by the main commercial airframe manufacturers and two commercial airlines.

---

[1]NRC, 1980, Evaluation of NASA's Program for Improving Aircraft Fuel Efficiency, OSTI ID 6589834, Washington, D.C, January 1.

[2]Richard T. Whitcomb, 1976, "A design approach and selected wind-tunnel results at high subsonic speeds for wing-tip mounted winglets," NASA TN D-8260.

[3]It should be noted that these results were obtained for a particular wing, winglet, and wingtip extension. Potential drag savings and moment distributions depend strongly on the geometry of the surfaces.

Airframe Manufacturers

Boeing

Boeing 7-Series aircraft have been manufactured by the "heritage" Boeing Commercial Airplanes Company in the Seattle area. The first in-production winglet produced by Boeing was for the 747-400, an improved Boeing 747 introduced in the late 1980s. The wingtip modification introduced on the 747-400 included a 6-ft per side wingtip extension and a canted 6-ft per side highly swept winglet (see Figure 3-1). The purpose of the wingtip modification was to improve the cruise aerodynamic efficiency of the airplane and, to a certain extent, differentiate the -400 model of the 747 family from the earlier -100/200/300 models. The tip modification increased the cruise L/D approximately 4 percent (less than half of the upper limit of 9 percent suggested by wind tunnel tests in the ACEE program, see above), with much of the improvement coming from the span extension. The L/D increase gives an equivalent increase in fuel efficiency.

Boeing introduced the 737 Next-Generation (NG) aircraft in the late 1990s without a tip device. Several years after entry into service, the Boeing

FIGURE 3-1 Boeing 747-400 with swept, canted winglets. SOURCE: Reproduced by permission of Boeing.

FIGURE 3-2 Boeing 737-NG with blended winglets. SOURCE: Reproduced by permission of Boeing.

Business Jets (BBJ) Company accepted a proposal to test a set of blended winglets[4] on the 737-BBJ from Aviation Partners Incorporated (API), a small Seattle-based aircraft modification company. The API design was a relatively large winglet, 8 ft long, installed on a relatively small commercial airliner. API required the intimate product knowledge of the original equipment manufacturer (OEM), Boeing, to be able to successfully integrate this winglet with the airplane structure and systems. In a joint flight development program with Boeing, the API winglet demonstrated a 4-5 percent block fuel reduction on the 737-NG series of aircraft (Figure 3-2).

The original design of the 737-NG wing allowed the winglet to be installed with only minor modifications to the wing structure. Owing to its nonstrength design features such as minimum gages required for hail and lightning protection, stiffness requirements for flutter, and design conservatism, the outer wing of the 737-NG had sufficient structural margins to

[4]Blended winglets reduce wetted area and junction drag.

accommodate the winglet readily. This made retrofitting the API winglets to the 737-NG technically and economically feasible. The BBJ market was a good trial for the winglets in that owners and operators of these airplanes wanted a high-tech look for their airplanes as much as they wanted efficiency benefits. Once the development work for the BBJ application had been completed, extending blended winglets to the commercial fleet of the 737-NG became a business decision that was accelerated by the rapid rise in fuel prices and economies of scale for large-volume production of winglets. Following the successful certification of the BBJ and commercial retrofit winglet design, Boeing then modified the in-production 737-NG design to accommodate the winglets, and that has become the almost de facto standard configuration for the 737-NG. Several thousand sets of 737-NG winglets had been ordered through the end of 2006 by the overwhelming majority of 737-NG operators.

Based on the success of the 737-NG winglet design, Aviation Partners Boeing (APB) was formed after Boeing purchased a minority interest in API. APB then developed a retrofit-only winglet installation for the Boeing 757 airplane (Figure 3-3). While the 757 is no longer in production, there are approximately 1,000 airplanes in service with the potential for another 20 years of life. APB was ultimately able to develop a retrofit package that uses the same winglet as the 737-NG on the 757. This was accomplished by developing a 17.5-in. tip extension that provides a transition from the 757 wingtip to the 737-NG wingtip, thus enabling a common interface to the existing winglet contour. Because of this tip extension and because the 757 wing did not provide as much excess structural margin as the 737-NG wing, the weight penalty of the 757 installation is considerably larger than that of the 737-NG. The total 757 installation weighs 1,358 lb versus approximately 340 lb for the 737-NG. However, the 757 tip extension also increases the efficiency gains for the installation, resulting in a block fuel savings potential of up to 5 percent, depending on mission range. A number of large domestic 757 operators, including American Airlines, Continental Airlines, and Northwest Airlines, have now opted for the APB winglet retrofit.

APB is also pursuing winglet retrofits for several other Boeing 7-Series aircraft. Recently, a retrofit winglet was certified for earlier models of the 737 family (737-300/400/500). This fleet went out of production in 1998, but there are still approximately 2,000 aircraft in service with decades of remaining life. APB is also investigating the feasibility of a winglet retrofit for the 767 family and the earlier models of the 777 family. An earlier pro-

FIGURE 3-3 Boeing 757 with retrofit blended winglets. SOURCE: Courtesy of Adrian Pingstone.

gram to develop an improved winglet installation for the 747-400 was not successful. The lesson from this experience is that not every airplane is a good candidate for further modifications. Success depends on aerodynamic compatibility and structural features such as strength margins and flutter margins.

Boeing took a different approach to improve the performance of its latest models of the 767 and 777 families. Both the 767-400ER and the 777-200LR/300ER incorporate a raked wingtip span extension design (Figure 3-4).[5] Similar in effect to winglets, the raked tips provide a reduction in cruise fuel burn and improved takeoff performance at the expense of longer wingspan. Boeing chose these designs because market studies indicated that the airplanes would still be able to use the same infrastructure as the older airplanes they would replace. The 767-400ER would be replacing DC-10 and L-1011 aircraft and competing with Airbus A330 aircraft, and the 777-200LR/300ER would be replacing or competing with Boeing 747-200/300/400 and Airbus A340 aircraft. In addition, the raked tip offers a

[5]Raked tips provide additional sweep.

FIGURE 3-4 Boeing 767-400ER with raked wingtips. SOURCE: Reproduced by permission of Boeing.

takeoff performance advantage over winglets because it improves not only drag but also lift, both of which are important for takeoff. Finally, the raked tip proved to be more efficient structurally because its design provides more aeroelastic relief than winglets for critical structural design conditions. The engineering trade-off for winglets versus raked tip extensions is a close call, and for these two aircraft families—767 and 777—the design space was more favorable for the raked tips than the winglets.

Boeing is currently designing a new family of long-range, wide-body transports, the 787 family (Figure 3-5). The 787 is intended to replace the 767 family and the Airbus A300/A330/A340 families. For reasons similar to the design solutions discussed for the 767-400ER and 777-200LR/300ER, Boeing selected a raked tip for the base 787-8 airplane. For growth versions of the 787, an even larger raked tip is envisioned, because the airplane will easily fit in the infrastructure that exists for the airplanes that it will replace. However, for the shorter range 787-3 model, which will be used for regional

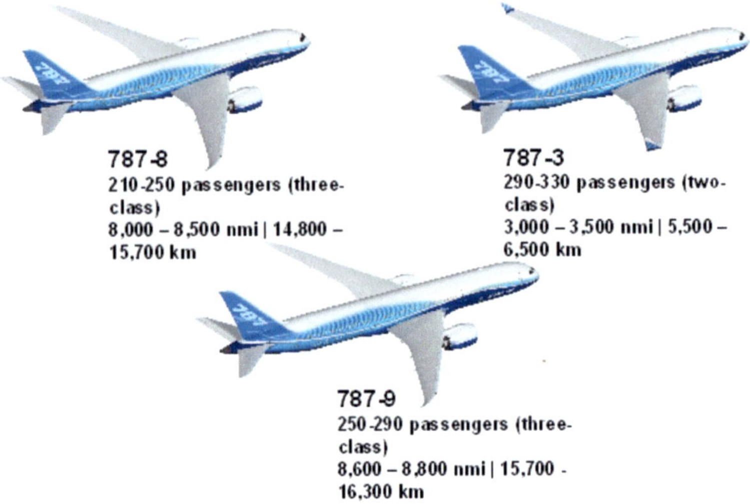

FIGURE 3-5 The Boeing 787 family, featuring various wingtip modifications. SOURCE: Reproduced by permission of Boeing.

transport to smaller airports, the market asked for a smaller span that would be more compatible with DC-10/L-1011/A300/767 operations. To satisfy that market, Boeing has shortened the span of the 787-3 wing and recovered some of the lost efficiency by developing a large vertical winglet. That design was still evolving at the end of 2006, and at this writing the precise outcome is still to be decided. The 787 strategy does provide an excellent example of the design trades that need to be made in order to decide which wingtip solution is appropriate for a given aircraft.

McDonnell Douglas Heritage Commercial Aircraft

Aircraft produced by the McDonnell Douglas Corporation, now part of Boeing, continue to be a large segment of the installed fleet of commercial airplanes. The two main families include the small twinjet DC-9/MD-80/MD-90/Boeing 717 models and the large trijet DC-10/MD-10/MD-11 models.

The trijet family has had numerous winglet studies conducted, including work done in cooperation with NASA in the 1970s and 1980s. The

FIGURE 3-6 McDonnell Douglas MD-11 with dual winglets. SOURCE: Reproduced by permission of Boeing.

DC-10 work in the early 1980s culminated in a flight test demonstration program that validated a 2-3 percent cruise efficiency improvement over the original DC-10 design, depending on the height of the winglet utilized.[6] However, a production winglet design was not incorporated until the design of the MD-11, a derivative of the DC-10 (the military KC-10 is also a derivative of the DC-10). The winglet configuration included a large upward-canted winglet and a small downward-canted winglet, as shown Figure 3-6. The selection of the dual winglet configuration was driven by the additional cruise drag benefit of the added span offered by the lower winglet and the favorable aerodynamic interactions between the upper and lower winglets at low-speed, high-angle-of-attack conditions. This configuration was developed in the early days of computational fluid dynamics (CFD) codes, when the original 747-400 winglet was being developed. Today's CFD capabilities

[6]A.B. Taylor, 1983, "DC-10 winglet flight evaluation summary report," NASA-CR-3748, December.

are much improved; the Navier-Stokes codes are considered capable of generating more accurate results and have been used in the latest aircraft designs to reach a successful design more quickly and with less wind tunnel testing.[7] In both cases the selected design philosophies might have been different had the current CFD capabilities and design lessons learned been available. Nevertheless, both the 747-400 and DC-10/MD-11 winglet designs have provided substantial airplane performance benefits to their products. The dual winglet design developed for the DC-10 has only been incorporated into the production MD-11 aircraft. Although the winglet was not flight tested separately from other aerodynamic modifications, it has been credited with a performance benefit of 2.5 percent.[8] To date, there has not been a retrofit program for the DC-10 aircraft.

The Douglas twinjet family has also been the subject of wingtip redesign studies. In the early 1980s, the DC-9 was redesigned as the MD-80, including a wing root insert and a wingtip extension. These changes provided more wing area, more wing span, and increased fuel volume, allowing increases in payload and range for this aircraft family. Also included in the MD-80 transformation were new, higher thrust/higher efficiency engines and an elongated fuselage. The wingtip extension for the MD-80 was notable in that it was a constant chord design, allowing the existing tip fairing and navigation light design to be retained. However, the wingtip extension and span loading were not optimized for efficient long-range cruise, as some other designs have been. The result of that is that the MD-80 wingtip devices do not show significant fuel economy benefits.

APB has investigated a retrofit design for the DC-9 family. Those design studies have not been successful in creating a viable business case. Projected block fuel burn reductions of less than 2 percent are offset by substantial modification costs. The limited potential for the DC-9 is a result of the existing wing structure, which hinders installation of a large winglet, as was possible on the Boeing 737 family. Since the DC-9 has been out of production since the early 1980s, the fleet size has shrunk and the fleet has aged, making the business case for a retrofit winglet or wingtip not as attractive as that for the Boeing 737 and 757 families.

---

[7]Forrester T. Johnson, Edward N. Tinoco, and N. Jong Yu, 2003. "Thirty years of development and application of CFD at Boeing Commercial Airplanes, Seattle," AIAA-2003-3439, 16th AIAA Computational Fluid Dynamics Conference, Orlando, Fla., June 23-26.

[8]Robb Gregg, Senior Manager for Aircraft Programs, Boeing Phantom Works, "Drag improvement: A study of the DC-10/MD-11/C-17 winglet programs," Presentation to the committee on December 13, 2006.

Just as Boeing and McDonnell Douglas were merging in 1997, a new derivative of the DC-9 family emerged as the renamed Boeing 717. That aircraft essentially combined the airframe of the original DC-9 with new engines and new systems. Since nearly 200 Boeing 717 aircraft were delivered before production terminated in 2006, there may still be a retrofit potential for this very new fleet. However, since the airframe is essentially a DC-9, it is unlikely that an outcome better than the projected reduction of 2 percent or less is possible without a significant structural modification of the aircraft. As of late 2006, no retrofit solutions for the twinjet family were being pursued.

Airbus Industries

There are two distinctly different winglet design strategies apparent on the commercial aircraft produced by Airbus Industries. The first is a "tip fence" concept, employed on the A310/A320/A380 families. The tip fence is a small dual winglet configuration with highly swept, nearly vertical upper and lower partial-chord winglets (Figure 3-7). For both the A310 and A320, the size of these winglets indicates that they were installed to take advantage of structural margin in the wings, since both aircraft were initially certified with plain wingtips.

A similar configuration was included in the initial rollout configuration of the A380, which was certified in December 2006. The design that preceded the A380, the design for the A330 and A340, had large, single-canted and highly swept winglets similar to the 747-400 configuration (Figure 3-8).

In 2006, Airbus flight tested several winglet designs on the A320 that appear to be similar to the 737-NG blended winglets. According to media reports, Airbus has decided not to offer these winglets for production or retrofit because the aerodynamic efficiency benefits determined from flight testing were not sufficient to overcome the very large weight increase needed for the installation. Airbus also expressed concern over potential long-term effects on the structural integrity of the wing due to stress imparted by winglet forces.[9] An implication to be drawn is that the wing may require significant structural modifications in order to accommodate the additional loads and/or flutter requirements of the large winglets. The Airbus experi-

[9]M. Kingsley-Jones, 2006, "Airbus rethinks plan to put winglets on A320," *Flight International*, October 10.

FIGURE 3-7 Airbus A320 with tip fence. SOURCE: Reproduced by permission of EADS North America.

FIGURE 3-8 Airbus A340 with swept, canted winglets. SOURCE: Reproduced by permission of EADS North America.

ence provides a valuable lesson on the difficulties that may be encountered in the design of winglets for retrofit to an existing aircraft.

Airlines

A number of airlines throughout the world have ordered or retrofitted some of their aircraft with winglets. The following summarizes the rationale and results for two commercial airlines that have chosen to modify their in-service aircraft with winglet devices.

Southwest Airlines

In 2000, when the price of oil reached $27 per barrel, Southwest Airlines conducted a study to retrofit a portion of its existing Boeing 737 fleet of aircraft with winglets as a means of reducing fuel burn. Although the proposed modifications held the potential for market expansion and increased revenue due to improved range and takeoff gross weight performance and other performance considerations, as well as the residual value of the modified aircraft, the original study payback period was based solely on the financial justification provided by reduced fuel burn.[10] The intent was to demonstrate the required performance benefits with no degradation in operational capability.[11]

The study of the 737-700 indicated a block fuel burn improvement (total fuel burn from engine start at the beginning of a flight to engine shutdown at the end of that flight) of 2.4 percent for flight segments of 500 nmi to a maximum of 4.0 percent for flight segments of 2,000 nmi. This improvement translated into a potential annual fuel savings of up to 110,000 gallons per aircraft for the modified fleet.[12]

With a projected payback period of 2 years for the first batch of modified 737 aircraft, the fuel burn savings satisfied financial and operational considerations for Southwest and the first order for winglets was placed in 2003. The modification process was coordinated through a Boeing Service Engineering team that operated as the single point of contact between

[10]Jim Sokol, Vice President of Maintenance and Engineering, Southwest Airlines, Conversation with the committee on December 14, 2006.

[11]Jay Inman, Vice President of Programs, APB, "Blended winglets," Presentation to the Committee on Analysis of Air Force Engine Efficiency Improvement Options for Large Non-fighter Aircraft on June 14, 2006.

[12]Ibid.

Southwest and APB in order to improve communications and program management. Southwest planned for a 7-day out-of-service time for each aircraft winglet modification, but its experience demonstrated that the modification could be accomplished in 3 or 4 days.[13]

In addition to the planned fuel savings, the modification demonstrated some increase in takeoff gross weight (TOGW) capability. As a result, the airline has benefited from an increase in stage lengths by adding takeoff fuel at facilities where the aircraft were previously limited by TOGW.[14]

The potential issues of ground damage, lightning strike, and hangar clearance that constituted the airline's internal justification for the modification program have proven to be of no consequence. Southwest reports that there have been no appreciable costs or operational limitations tied to crew training, technical data, or the like.

The success of the 737-700 modification program has motivated Southwest to initiate a new proposal for older 737-300 aircraft that lack a suitable wing structure. The extra weight for this modification is estimated to be 783 to 801 lb. This proposal is based on a projected block fuel burn improvement of 2.6 percent for a 500 nmi stage and 4.4 percent for a 2,000 nmi stage and could save up to 100,000 gallons of fuel per aircraft per year.[15] Because the wing structure must be modified, the out-of-service time to complete the winglet modification for the 737-300 aircraft is estimated to be 14 days.[16] The results of the Southwest 737 winglet modification program are summarized in Table 3-1.

---

[13]Jim Sokol, Vice President of Maintenance and Engineering, Southwest Airlines, Conversation with the committee on December 14, 2006.

[14]Ibid.

[15]Jay Inman, Vice President of Programs, APB, "Blended winglets," Presentation to the Committee on Analysis of Air Force Engine Efficiency Improvement Options for Large Non-fighter Aircraft on June 14, 2006.

[16]Jim Sokol, Vice President of Maintenance and Engineering, Southwest Airlines, Conversation with the committee on December 14, 2006.

TABLE 3-1 Southwest Airlines 737 Winglet Modification Summary

Aircraft Type	Retrofit Weight Increase (lb)[a]	Production Weight Increase (lb)[b]	Increase in Wing Dimensions Attributable to Winglets	Block Fuel Savings (%)
737-700 (non-provisioned wing)	340	N/A[c]	8 ft 2 in. (height) 6 ft 4 in. (span)	2.4 to 4.0
737-700 (provisioned wing)	241	220	8 ft 2 in. (height) 6 ft 4 in. (span)	2.4 to 4.0
737-300	783 to 801	N/A[c]	7 ft 6 in. (height) 8 ft 10 in. (span)	2.6 to 4.4

[a]Difference between the weight of the aircraft manufactured without winglets and the weight of the aircraft after retrofitting with winglets.

[b]Difference between the weight of the aircraft manufactured without winglets and the weight of the aircraft with winglets.

[c]Aircraft are not manufactured with winglets as original equipment.

SOURCE: Data from APB.

American Airlines

Following an extensive study, American Airlines decided in 2004 to add winglets to its long-range international fleet of 20 757-200ER aircraft.[17] The decision was made subsequent to a detailed business planning effort. American considered the following costs and benefits in its winglets business case:[18]

The business case for installing winglets considers the following costs:

- Materials (hardware, software, consumables),
- Labor,
- Tooling,
- Spares for inventory,
- Expected maintenance (repairs, inspection, replacements),
- Out of service (lost contribution),

[17]John Novelli, Director, Operations Engineering and Optimization, American Airlines, "American Airlines winglets," Presentation to the committee on December 13, 2006.

[18]Ibid.

- Ferry costs,
- Fuel (use forward-fuel-curve-based futures market),
- Airport gating loss, modifications, or flexibility costs, and
- Maintenance hangar modifications at bases and airports.

The business case for installing winglets considers the following benefits:

- Improved fuel burn,
- Engine derate,
- Possible new markets due to increased range,
- Additional payload, and
- Change in residual value of the aircraft.

Other factors for consideration:

- Lease agreements,
- Coordinating modifications with other base maintenance visits, and
- Economic life of aircraft.

American found that the winglet was supported on the basis of fuel conservation alone if a high NPV discount rate was used,[19] and it contracted with APB to manufacture the winglets. The negotiation yielded a single extra kit to install on one of American's 737-800s as a test case for the benefits on that fleet of aircraft. Prior to the modification program, American's 737-800 fleet was operating at an average +2.2 percent fuel burn over the specification, or book, level (worse than design level).[20] The modification program had the potential to return fuel burn to book level or better. The first winglets were installed on the 737-800, and the modified aircraft became operational in October 2005.

Whether the winglets are original equipment or retrofitted, the weight penalty for the aircraft is 380 lb. As with Southwest's installation on its 737-700s, the vertical portion of the winglet adds 8 ft 2 in. to the height of the wing, while the horizontal portion of the winglet modification adds 4 ft 7 in. to the wingspan.

---

[19]Ibid.

[20]Ibid.

The modified 757-200ER aircraft was released to service in December 2005. The weight added by the winglet modification was approximately 1,400 lb. To mitigate the added weight, American obtained an increase in the certified maximum zero fuel weight. The effect on wing geometry for the winglet modification is an increase in wing height of 8 ft 2 in. and an increase in wingspan of 9 ft 9 in.

American used two separate methods (actual flight burn data and aircraft performance monitoring software) to calculate the fuel savings realized by the winglet modification of its 737 and its 757 aircraft. Those calculations demonstrate a fuel savings for the 737-800 aircraft of 3.2 percent when compared to its 737-800 nonwinglet fleet. This equated to 32 gal of fuel saved per flight hour.[21]

The modified 757-200ER fleet demonstrated a fuel savings of 3.3 percent in comparison with its unmodified 757-200ER fleet, a fuel savings of 40 gal per flight hour.[22] Both findings are in line with reports of savings from other operators of similar aircraft.[23]

The results of American's 737-800 and 757-200ER winglet modification programs are summarized in Table 3-2.

American concluded that there were no changes in flying qualities, and the Federal Aviation Administration (FAA) required no changes to the flight simulators of either fleet as a result of the winglet modifications. The FAA deemed that the flight crew training requirements could be satisfied with appropriate technical manuals.

The added winglets improved the takeoff performance of American's aircraft and gave them greater takeoff gross weight capability. These resulting improvements expanded the airline's market potential by giving it access to previously climb-limited airports (generally high-altitude, high-temperature, or short-field airports). American has taken advantage of these new capabilities and thereby increased its revenue.[24]

---

[21]Ibid.
[22]Ibid.
[23]Ibid.
[24]Ibid.

TABLE 3-2 American Airlines 737-800 and 757-200ER Winglet Modification Summary

Aircraft Type	Retrofit Weight Increase (lb)[a]	Production Weight Increase (lb)[b]	Increase in Wing Dimensions Attributable to Winglets	Block Fuel Savings (%)
737-800 (non-provisioned wing	520	N/A[c]	8 ft 2 in. (height) 4 ft 7 in. (span)	3.2
737-800 (provisioned wing)	380	380	8 ft 2 in. (height) 4 ft 7 in. (span)	3.2
757-200ER	1,358	N/A[c]	8 ft 2 in. (height) 9 ft 9 in. (span)	3.3

[a]Difference between the weight of the aircraft manufactured without winglets and the weight of the aircraft after retrofitting with winglets.

[b]Difference between the weight of the aircraft manufactured without winglets and the weight of the aircraft with winglets.

[c]Aircraft are not manufactured with winglets as original equipment.

SOURCE: Data provided by APB and American Airlines.

MILITARY EXPERIENCE

C-17

The C-17 was designed by the McDonnell Douglas Corporation, now a part of the Integrated Defense Systems component of the Boeing Company (Figure 3-9). Winglets were incorporated into the design for reasons relating to taxi clearance, turning radius, maneuverability, and parking. In particular, the Air Force wanted to limit the wingspan to that of the C-141 to make the C-17 compatible with facility infrastructure. Clearly, it was preferable to achieve the desired airplane cruise performance by adding winglets rather than increasing the wingspan.

A number of these design considerations led to the final winglet configuration on the C-17. Early C-17 designs included upper and lower winglets. However, the lower winglet was eliminated after it was determined that the cruise performance goals could be met with a single upper winglet configuration; that ground clearance requirements would be problematic with the lower winglet; and that the lower winglet would result in higher manufacturing and maintenance costs. The planform and placement of the C-17 winglet were also driven by exterior lighting requirements.

FIGURE 3-9 The Air Force's C-17 with winglets.

The winglet was shown in wind tunnel testing to reduce cruise drag approximately 2.5 percent. Also, no additional buffeting was observed for takeoff or landing configurations during flight testing.[25] However, while these benefits are considered to be substantial, the C-17 winglet was developed in a low Reynolds number (RN) wind tunnel. The low RN environment can produce misleading results with regard to drag, buffet, pitching moment, and loads because the flow phenomena are different from those at the much higher RN of the full-scale flight vehicle. Also, the C-17 configuration was developed in the 1980s before the full-scale National Transonic Facility Wind Tunnel and before modern Navier-Stokes CFD tools. With these new capabilities, the current C-17 winglet design could be more accurately assessed. These new tools, together with lessons learned from other winglet designs, might make it possible to improve the C-17 winglet design, and thereby cruise drag, another 1 percent or more.[26]

---

[25]Robb Gregg, Senior Manager for Aircraft Programs, Boeing Phantom Works, "Drag improvement: A study of the DC-10/MD-11/C-17 winglet programs," Presentation to the committee on December 13, 2006.

[26]Ibid.

C-32

The C-32 is a military derivative of the Boeing 757-200 commercial aircraft and is used for government executive transport. APB offers a retrofit package for this aircraft, and a contract was awarded in October 2006 for installation of this package on the four C-32 aircraft in the Air Force inventory.

KC-135

A number of studies have been conducted over the years to determine the suitability of adding winglets to the KC-135 aircraft, which is closely related to the Boeing 707. Some of the pioneering work on winglets was conducted at NASA in the mid-1970s on the KC-135. That work was followed by several Air Force contracts with Boeing to investigate the design space for winglets on the KC-135 and included extensive wind tunnel testing. That work, in turn, determined that winglets could greatly improve cruise efficiency. An improvement in the L/D of nearly 8.5 percent was reported, along with an estimated empty weight increase of approximately 600 lb, for a net performance improvement of nearly 7.5 percent.[27] The winglet selected for this work was nearly 9 ft in length, slightly larger than the production winglets on the 737-NG.

The improvement predicted for this winglet on the KC-135 is considerably larger than that for any winglet that has actually been incorporated into an airplane. In the 1970s, it was speculated that the KC-135 would be an excellent candidate for winglets because the wing was overloaded relative to the ideal elliptical span load, which would presumably allow the winglet to load more optimally. While that explanation may have had some merit, it would not be surprising if the current methods used to design and analyze winglets were to arrive at a less optimistic prediction. Certainly it is now known that the aeroelastic impact of the winglet on the wing's twist must be considered, and that should decrease the benefit somewhat. Nevertheless, the work conducted in the 1970s suggests that a significant benefit could accrue from using winglets on the KC-135 and that the structural requirements for installation appear to be manageable.

Later the Air Force conducted a flight test of winglets on a KC-135. The published report on that flight test indicated good agreement with

---

[27]K.K. Ishimitsu, 1976, "Design and analysis of winglets for military aircraft," AFFDL-TR-76-8, Wright-Patterson Air Force Base, Ohio, Air Force Dynamics Laboratory.

the analytical and wind tunnel testing when known differences between the tests were accounted for.[28]

SUMMARY OF COMMERCIAL AND MILITARY EXPERIENCE

The commercial airframe manufacturers began testing winglet installations for several aircraft more than 20 years ago. At the time many of these ideas were first proposed, fuel prices were not sufficiently high to justify retrofit costs. However, the retrofit idea took off once fuel prices started rising, and airlines saw the need to use more economical aircraft for longer-range missions. Cost-benefit analyses conducted by the airlines (American Airlines and Southwest Airlines) indicated that winglet retrofit programs on appropriate aircraft would pay for themselves within about 2 years based on the fuel savings alone.

The Boeing 737-NG became the flagship for the retrofitting of commercial fleets with winglets. It was an aircraft that was well suited since it needed very little structural upgrade, was easily modifiable, and proved to have 3-5 percent fuel burn improvements depending on mission length and other factors. The installation takes only 4 or 5 days, and the aircraft have had no negative operational issues nor have they needed any change to flight operational procedures or training of crew.

The winglet-modified Boeing 757 shows similar fuel burn improvements, and this makes it ideally suited for secondary European markets from the U.S. East Coast. It actually has allowed using a smaller aircraft that was originally intended for domestic use, turning it into an efficient international aircraft.

The commercial experience is that wingtip modifications make sense if one can achieve a 3-5 percent fuel burn improvement, if careful engineering analysis shows that the aircraft have sufficient structural integrity to easily accept wingtip extensions or winglets, and if the modifications are relatively easy to install. The airlines have been able to overcome with little difficulty the initial concerns relating to the added wing height and wingspan in hangars, at gates, and on taxiways.

Only one military-unique aircraft, the C-17, features winglets. Designers had a choice of either increasing the wingspan or using winglets to achieve the desired performance, and winglets were chosen because they

---

[28]NASA, 1982, "KC-135 winglet program review," NASA Conference Publication 2111, January.

minimize problems relating to taxi clearance, turning radius, maneuverability, and parking. However, the C-17 design was done before modern analysis and optimization tools were fully developed, and application of these tools could further improve the C-17's aerodynamic performance.

As discussed earlier, the retrofit potential of some other military aircraft, such as the KC-10 (based on the DC-10 airframe) and the KC-135 (which is closely related to the Boeing 707 airframe), has been studied and found promising. Other military-unique aircraft, such as the C-5, would require extensive engineering analysis before a judgment could be made. In Chapter 4, the committee reviews all the candidate aircraft in the Air Force inventory and recommends those that merit careful consideration for tip device retrofits.

4

Assessment of Wingtip Modifications for Various Air Force Aircraft and Potential Investment Strategies

This chapter provides the committee's evaluation of steps the Air Force could take to improve the fuel economy of aircraft in its inventory, in particular by modifying the wingtips. It begins with a checklist of factors that must be considered to determine if these modifications make sense. This is followed by a discussion of specific aircraft in the Air Force inventory, including those that are responsible for the greatest fuel consumption as well as those that are derived from commercial aircraft. The committee then identifies those aircraft that appear most promising for wingtip modification. For these selected aircraft, a simple spreadsheet model is used to estimate payback periods for modification investments, treating modification costs and fuel prices as parameters. These calculations are combined with the committee's expert judgment to prioritize various aircraft for their suitability for wingtip modifications. Finally, the chapter concludes with a discussion of strategies by which the Air Force can optimize its investment in fuel economy programs.

CHECKLIST FOR MAKING WINGTIP MODIFICATION DECISIONS

The investment in winglets for a particular aircraft type depends on a number of factors, including the potential fuel burn efficiency improvements provided, the size of the statement of work required for the installation, the utilization rate of the aircraft fleet, and the expected lifespan of that particular fleet. An extensive engineering and economic analysis would be required for each aircraft type in order to determine the appropriateness of installing winglets. The following elements are necessary in order to make a balanced decision for each aircraft fleet.

Technical Issues

Cruise Fuel Burn Efficiency Improvement

The primary reason for installing winglets (or other tip devices) is to improve the efficiency of the fuel burn at cruise conditions of the aircraft. The two most important components of fuel burn efficiency affected by winglets are the aerodynamic efficiency of the configuration, measured in terms of lift-to-drag ratio (L/D), and the empty weight of the aircraft, which will increase when the winglets are installed. The viability of a winglet installation is different for each aircraft configuration, and sophisticated design studies are required to achieve the proper balance between aerodynamic efficiency and weight efficiency. There are numerous design parameters involved in selecting the optimum winglet configuration, including winglet span, area, sweep, taper, cant angle (inclination), twist, thickness, sweep, juncture flow, etc. The selection of materials for winglet construction will affect the empty weight. Moreover, the additional loads and moments imparted to the wing when the winglet is installed may require that the wing be strengthened, adding more weight. A sophisticated dynamic aeroelastic analysis of the wing/winglet structure is required for this assessment.

Collateral Impact of the Winglet Installation on Airplane Design

In addition to the aerodynamic and structural effects of the installed winglet, ancillary issues related to the winglets must also be considered, including the need to revise flight control systems, brought about by the changed stability and control characteristics. These include changed longitudinal, lateral, and directional stability characteristics and altered control system effectiveness, particularly with regard to the effectiveness of outboard ailerons and spoilers. Winglets also can affect the configuration of tip lighting systems and the lightning strike protection systems for the wing.

Collateral Impact of the Winglet Installation on the Infrastructure

The interaction of the airplane with its infrastructure must also be factored into a winglet decision. Typically the physical span of the aircraft increases with the installation of winglets, but not as much as with a conventional tip extension. Nevertheless, consideration must be given to issues related to ground handling, parking, and maintenance (depot and field) and associated facilities such as gates, ramps, hangars, runways, taxiways, etc. This is particularly important when analyzing the economic life cycle of winglets.

Design Information Availability

Developing a winglet design for an existing aircraft requires a deep understanding of the characteristics of the original aircraft design. Generally that detailed design knowledge resides primarily with the original equipment manufacturer (OEM). However, there have been successful retrofit designs of winglets that were originated by third-party companies. For older aircraft, the existing design data may be scarce and not compatible with current design tools. In addition, there may be few, if any, engineering personnel with a working knowledge of that particular aircraft design. These factors must be considered in developing a financial estimate for the cost and risk of developing a winglet retrofit design.

Economic Issues

In addition to the formidable technical challenges of developing a winglet retrofit configuration, there are significant economic factors that come into play when making a life-cycle business case. Among the factors that must be considered are the following:

- *Cost of installation.* A contractor will need to charge a reasonable price to establish a positive business case for proceeding. The fixed cost to the contractor will consist of engineering and tooling costs required to design, test, and validate the winglet configuration. That cost is nearly independent of the size of the fleet of airplanes, so the larger the subject fleet, the more units that the fixed design costs can be amortized over, making the business case for winglets more likely to be favorable for a large fleet of aircraft.

- *Life span of the fleet.* A retrofit design solution will have a potentially longer payback period for a younger fleet of airplanes with a longer economic life than for an aging fleet that is soon to be retired. This economic factor can also be influenced by the rate at which the retrofit is conducted. A slow retrofit program eats into the payback on the initial investment, while a rapid fleet retrofit accelerates the payback period.
- *Utilization rate of the fleet.* Winglets reduce the fuel burn per flying hour of an aircraft. The more the aircraft is used, the faster the investment will be paid back. This favors installing winglets on heavily used fleets.
- *Cost of fuel.* Since the means of paying back the initial investment is a reduction in the amount of fuel consumed, costlier fuel means that the payback is quicker and more likely. Less costly fuel requires a longer payback period.
- *Cost of capital.* As with any up-front investment, there is a cost for the capital that is expended before payback can occur. Assuming that the capital investment is made with borrowed money, the economic environment in terms of interest rates and inflation must be considered to understand the business case. High interest rates and low inflation will adversely affect the business case, while low interest rates and high inflation will make the cost of borrowing less.

Putting It All Together

A business case model can be created to establish the viability of a winglet retrofit program for a fleet of airplanes. Independent variables in the assessment include the following:

- Winglet unit price ($/airplane),
- Fuel burn reduction (%),
- Cost of fuel ($/gallon or lb),
- Interest rate (%),
- Inflation rate (%),
- Fleet size (number of airplanes),
- Fleet utilization (hours/year),
- Retrofit rate (airplanes/year), and
- Life span remaining (years).

These variables can be used in a business case model to determine the cash flow profile. The profile will be negative during the development and early retrofit years and should become positive during the lifetime of the program. If there is not a positive outcome, winglets should not be installed. If the outcome is positive but requires a long period to break even, the decision is not clear cut. If the outcome is positive over a short period, winglets should clearly be installed.

CANDIDATE AIRCRAFT IN THE AIR FORCE INVENTORY

Given the emphasis on fuel economy in the study's statement of task, the committee began by considering those aircraft that consume the greatest amount of fuel, as shown in Figure 4-1. The five that stand out most clearly are, in order of annual fuel usage by fleet, the C-17, KC-135R/T, C-5, KC-10, and C-130H/J. As noted in Chapter 3, the C-17 already has winglets, and the KC-135 and KC-10, which are closely related to the Boeing 707 and DC-10 commercial airframes, respectively, have been studied previously for wingtip modifications. The aircraft are discussed further below.

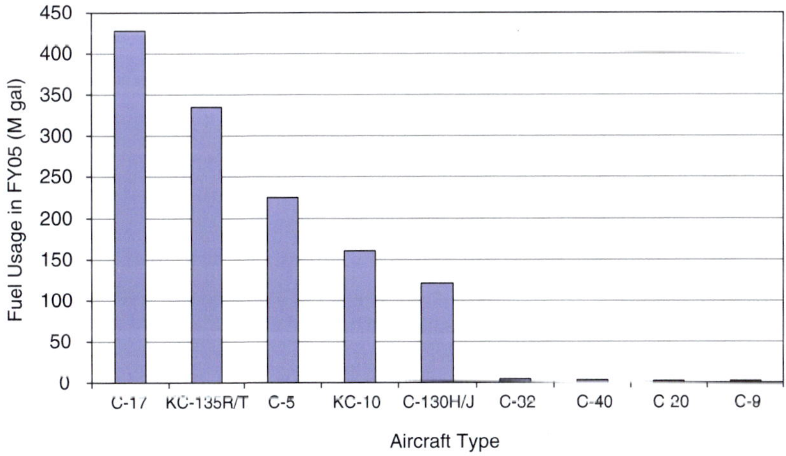

FIGURE 4-1 Fuel usage of selected Air Force aircraft (by fleet) in FY05. SOURCE: DESC.

C-17

The C-17 is the most current freighter aircraft and one that has some of the latest structural and aerodynamic improvements installed. As described in Chapter 3, the C-17 aircraft is already equipped with a winglet that was incorporated into the original design. While a newly designed winglet for the C-17 might result in somewhat improved cruise fuel efficiency, the magnitude of that improvement is likely to be in the 1 to 2 percent range,[1] and it would only make sense if combined with other efficiency improvement modifications. The considerable data already developed for the C-17 could also be considered for further wing upgrades beyond winglets. These should be reviewed and considered for possible installation since much of the research has already been accomplished.

KC-135R/T

As noted in Chapter 3, there has been some testing with winglets and other improvements on the KC-135/707 wings. These studies should be reviewed for applicability. The issue with these airframes is primarily their age and the limited remaining useful life. The current fleet of approximately 417 KC-135R/T aircraft would be good candidates for winglet installation. Some of these aircraft are expected to be in service until approximately 2040. The fleet of TF-33/JT-3D equipped KC-135s (D/Es) is potentially subject to retirement, so they may not have sufficient payback life remaining. The RC-135V/W/S/TC fleet may be candidates for winglets as well. However, the installation of the overwing high-frequency antenna and wingtip pitot-static probes would probably create another problem for the addition of winglets. Therefore, the committee believes the focus within the KC-135 fleet should be on the R/T models.

Three related military fleets derived from the Boeing 707 commercial aircraft are the E-3 AWACS, the E-6 TACAMO, and the E-8 JSTARS. While the wings of these aircraft are closely related to those of the KC-135 fleet, any such winglet would have to be further investigated to confirm aerodynamic and structural compatibility. In addition, there would need to be consideration of winglet interference with the AWACS antenna function. The TACAMO has an extended wingtip that houses antenna pods, so it is

---

[1]Robb Gregg, Senior Manager for Aircraft Programs, Boeing Phantom Works, "Drag improvement: A study of the DC-10/MD-11/C-17 winglet programs," Presentation to the committee on December 13, 2006.

very unlikely that those aircraft could be modified for winglets. Moreover, less is known about the structural suitability of these 707-based platforms than is currently known about the KC-135 fleet.

The KC-135/Boeing 707 aircraft are similar in gross weight to the Boeing 757 commercial aircraft, with a maximum takeoff weight (MTOW) of approximately 250,000 lb. The Boeing 737-NG winglet solution has been installed on the Boeing 757 using a 17.5-in. transition wingtip that accommodates the tip airfoil differences between the 737-NG and the 757. This experience may provide a solution for the KC-135/Boeing 707 military fleet as well.

C-5

The Lockheed Martin C-5 is a global strategic airlift system capable of carrying outsized cargo. A total of 111 C-5s are in service with the Air Force. A portion of the fleet is being modified with modern commercial turbofan engines, improving range by up to 11 percent. Aerodynamic range improvement efforts have focused on airframe housekeeping such as orphan weight removal and airframe cleanup. Given that the C-5 is one of the major contributors to fuel consumption by the Air Force and its missions are long range, a study to quantify the potential performance gains and integration effects of winglets is warranted. Unlike efforts devoted to winglets for the commercial transport, there has been no detailed C-5 winglet development effort, adding a sizable nonrecurring cost to a fleet retrofit program and extending the time required to recover the investment.

Specific data for C-5 aerodynamic improvements have not been approved for release by the System Program Office. The committee believes that the C-5 has the potential for drag reduction with wingtip modifications because of its current large fuel consumption, its missions, wing design, etc. Aerodynamic improvements, combined with orphan weight and obsolete component removal, would contribute to operating efficiency increases for the aircraft.

KC-10

The KC-10 is a military derivative of the McDonnell Douglas DC-10 commercial aircraft. As mentioned in Chapter 3, while there has not been a winglet retrofit program for the DC-10, winglets were successfully flight tested on the aircraft and were later successfully incorporated into the derivative MD-11 aircraft. There are currently 59 KC-10 aircraft in the Air Force inventory, and they are flown extensively. The KC-10 fleet is quite young, with the oldest aircraft having approximately 22,000 flight hours and 14,000 flight cycles. For comparison, the older DC-10 still in service had over 131,000 flight hours and 45,000 flight cycles as of September 2006. There are also approximately 150 of the DC-10 family of aircraft still in commercial service, both passenger and cargo, so the combined potential market for retrofits may be large enough to motivate a retrofit program.

The MD-11 experience indicates that a successful winglet can be incorporated into the DC-10-based wing design. The DC-10 flight test program that was conducted identified approximately 3 percent cruise efficiency improvement, which was later replicated on the MD-11 design.[2] In addition, recent winglet design experience using modern CFD methods and high Reynolds number (RN) wind tunnels may provide lessons that could have applicability for winglet designs that may be more effective on the KC-10 and other government transport aircraft. With new multidisciplinary design and optimization methods available, it is likely that an even more efficient and simpler design could be feasible for this aircraft family.

C-130H/J

The Lockheed Martin C-130 is a tactical airlifter designed to operate from short, austere airfields. While the committee's focus was on the C-130H/J, a total of 655 aircraft in 16 variations carry out a broad spectrum of missions, from intertheater airlift to electronic and psychological warfare.

In evaluating the suitability of the C-130 for the application of winglets to increase cruise efficiency, several factors suggest that performance might improve less than seen on commercial aircraft. The C-130's wing is already very efficient because its aspect ratio of 10 is relatively high, reducing the overall benefit expected from winglets. The wing design was driven by the need for short-field performance—a requirement not imposed on jet airliners—as well as cruise performance, resulting in the high-aspect-ratio geometry and associated high aerodynamic efficiency. A further reduction

[2]Carl A. Shollenberger, John W. Humphreys, Frank S. Heiberger, and Robert M. Pearson, 1983. "Results of winglet development studies for DC-10 derivatives," NASA-CR-3677, March.

in winglet effectiveness is attributed to the C-130's unswept wing with its lower tip loading.

Operational considerations also reduce the effectiveness of a winglet modification program. The C-130 missions tend to be short range and flown at lower altitudes. Since winglets are more effective for longer ranges and with higher wingtip loading (realized at higher altitudes), the potential benefit of winglets for the C-130 is limited.

A development effort would be needed to optimize winglet geometry, determine integration effects, and evaluate system-level benefits. Other drag reduction approaches, such as aft body strakes and revised wing fillets, have been identified in other studies and should be considered in any fuel consumption reduction study.

Intelligence, Surveillance, and Reconnaissance Aircraft

The committee was asked explicitly to consider the suitability of ISR aircraft for wingtip modifications. While the U-2 and Global Hawk fall into the ISR category, their mission requirements (extremely high altitude and long endurance) result in wing designs that are already extremely efficient and would be expected to show little if any benefit from winglets. In fact, there might be performance penalties for integrating winglets on these platforms because performance at high altitudes is extremely sensitive to weight. Thus, these aircraft are not good candidates for wingtip modification.

Other Air Force Aircraft

The easiest decisions on whether to install winglets obviously pertain to aircraft in the Air Force inventory that derive from commercial aircraft now operating with winglets. In each case, the aircraft structure has already been studied and found to be appropriate, the engineering design has been done, the modifications have been prototyped, tested, and certified, modification kits developed, flight manuals revised as required, and so on. However, the committee's review of all such Air Force aircraft revealed that most of them already have winglets or the decision has been made to incorporate winglets, as shown below in Table 4-1.

All of these aircraft have winglets except for the C-9s, the C-21s, the VC-25s, and the E-4s. The three C-9s, all derivatives of the DC-9, are scheduled to retire in FY11 and should not be considered for wingtip modifications. Also, past work on winglets for the DC-9, as discussed in

TABLE 4-1 Winglet Status of Air Force Aircraft Derived from Commercial Airframes

Air Force Aircraft	Commercial Equivalent	Inventory	Winglets
C-9	Douglas DC-9-30	3	No
C-20B	Gulfstream GIII	5	Yes
C-20H	Gulfstream GIV, GIVSP	2	Yes
C-37	Gulfstream GV	9	Yes
C-21	Learjet 35A	59	No
C-40B	Boeing 737-700	4	Yes
C-40C	Boeing 737-700	3	Yes
VC-25	Boeing 747-200 (-300 wings)	2	No
E-4	Boeing 747-200	4	No
C-32	Boeing 757-200	4	Yes

SOURCE: Data courtesy of U.S. Air Force.

Chapter 3, did not prove to be favorable. The C-21s, derivatives of the Learjet 35A, are small aircraft, and the entire fleet uses approximately 8 million gallons of fuel per year and would not be a priority to modify. Furthermore, they have tip tanks, and wingtip modifications would require the removal of these tanks, severely limiting the range of the aircraft even with a more efficient wing. Lastly, the VC-25s and the E-4s are derivatives of the Boeing 747-200, with the VC-25s having 747-300 wings. The 747-200 has not been produced since the late 1980s, so the commercial fleet is aging and retiring from service. As a result, the entire cost of winglets designed for 747-200/300 wings would have to be borne by the government. All of the 747s in the commercial world that have winglets are 747-400s, which have a structurally modified wing. The structural modification to allow installing the 747-400 wingtip on the VC-25s or the E-4s would be very expensive and impractical.

This discussion leads to the following finding:

Finding: Most of the aircraft in the Air Force inventory that derive from commercial aircraft now operating with winglets already have winglets, or the decision has been made to install winglets. The remaining Air Force aircraft that are derivatives of commercial aircraft do not appear to be good candidates for wingtip modifications.

TABLE 4-2 Potential for Wingtip Modifications to Benefit Air Force Aircraft

Aircraft	Priority/Potential Benefit
KC-10	High
KC-135R/T	High
C-5	Medium
C-17	Medium/low
C-130H/J	Low

PRIORITY AIRCRAFT TO BE CONSIDERED FOR WINGTIP MODIFICATION

Based on the committee's judgment on a variety of factors, some of which are detailed in the following pages, five aircraft were ranked in the order of their suitability for wing modifications, as shown in Table 4-2.

However, these judgments are based on minimal basic data, and a detailed engineering and economic analysis would be required for each aircraft type before a final decision could be made to proceed with the installation of winglets or other aerodynamic modifications.

PRELIMINARY NET PRESENT VALUE ANALYSIS

To illustrate the types of benefits and costs that might be realized through wingtip modifications (e.g., winglets) that would produce a reduction in fuel burn, the committee shows here, as examples, the results of its preliminary net present value (NPV) analysis for the KC-135R/T and the KC-10. Appendix A shows the sets of data values the committee used for both aircraft, including number of aircraft, fuel burn, flying hours, and projected retirement dates to calculate the NPV of savings. The mission profiles are inherent in the data used for each aircraft. In particular, the fuel consumption rates (pounds or gallons per hour) are the average over the various mission profiles actually flown.

Since it is not possible to know the fuel savings and modification cost for a specific aircraft without performing a detailed engineering analysis, as described earlier in this chapter, the committee parameterized fuel savings and modification cost for each aircraft.

The calculations were done for block fuel savings of 3 percent and 5 percent, consistent with commercial airline experience and the findings

TABLE 4-3 Estimated Aircraft Modification Costs[a]

Aircraft	Estimated Modification Cost Range (million $)
KC-135R/T	0.5-1.0
KC-10	1.5-3.0

[a]Includes nonrecurring development costs.

of this report. The price per gallon of fuel was parameterized at $2.50, $5.00, $10.00, and $20.00 to represent the fully burdened cost of fuel. (All monetary values are in dollars of 2007 purchasing power.)

The committee estimated one modification cost range for the KC-135R/T and one for the KC-10, as shown in Table 4-3.

For the NPV calculations, the committee assumed an annual fuel cost escalation rate of 3 percent and a discount rate of 3 percent.

Using the above costs and fuel saving and the data in Appendix A, the committee first calculated the time required for fuel savings to pay back the cost of modifying an individual aircraft.

The results shown in Tables 4-4 and 4-5 suggest that modifying the KC-135R/T and KC-10 aircraft in its inventory might financially benefit the Air Force.[3] Even in the worst case (highest modification cost, lowest fuel usage reduction, and fuel cost of $2.50 per gallon) for each, the payback periods are within the expected remaining service lives of the aircraft. The results also show how the payback period is affected by the cost of fuel. In constant dollars, if the cost of fuel were to double, the payback period would be cut in half.

The NPV results are shown in Figure 4-2 for the KC-135R/T and in Figure 4-3 for the KC-10. The figures show the estimated cumulative fleet net savings over time from the start of aircraft modification to when the last aircraft is retired from service. Results are shown for the worst-case (highest modification cost and lowest fuel usage reduction) and best-case (lowest modification cost and highest fuel usage reduction) payback periods at a fuel cost of $2.50 per gallon. These calculations also take into account the modification cost, aircraft-specific information such as number of

[3]The committee's parametric analysis suggests—but does not prove—that financial benefits would accrue from modifying these aircraft. As stated earlier in the report, deeper aircraft-specific engineering analysis is required to support more precise and higher confidence estimates of the costs and benefits of making the modifications.

TABLE 4-4 Payback Period for a KC-135R/T Using 649,000 gal/yr

Estimated Cost of Modification (FY07 $M)	Fuel Usage Reduction from Modification (%)	Fuel Saved (K gal/yr)	Fuel Cost Saved (FY07 $K)	Payback Period (years)
Fuel at $2.50/gal				
0.5	5	32	81	6.2
0.5	3	19	49	10.3
1.0	5	32	81	12.3
1.0	3	19	49	20.6
Fuel at $5.00/gal				
0.5	5	32	162	3.1
0.5	3	19	97	5.1
1.0	5	32	162	6.2
1.0	3	19	97	10.3
Fuel at $10.00/gal				
0.5	5	32	324	1.5
0.5	3	19	195	2.6
1.0	5	32	324	3.1
1.0	3	19	195	5.1
Fuel at $20.00/gal				
0.5	5	32	649	0.8
0.5	3	19	389	1.3
1.0	5	32	649	1.5
1.0	3	19	389	2.6

aircraft, projected lifetime, flight hours per year, and fuel burn. For these illustrative calculations, it was assumed that the nonrecurring engineering would be done by FY08 and the modifications would begin in 2009. The modifications would be done while an aircraft is undergoing regular depot maintenance, so it would not be out of service for any additional time. The committee also assumed for these calculations that all of the aircraft in the fleet would undergo programmed depot maintenance at a uniform rate between FY09 and FY13 inclusively.

As shown in Figure 4-2, the KC-135R/T best case, net savings become positive 9 years after starting the modification program. All 417 aircraft in the inventory are modified. Total net savings to the Air Force are approxi-

TABLE 4-5 Payback Period for a KC-10 Using 2.057 million gal/yr

Estimated Cost of Modification (FY07 $M)	Fuel Usage Reduction from Modification (%)	Fuel Saved (K gal/yr)	Fuel Cost Saved (FY07 $K)	Payback Period (years)
Fuel at $2.50/gal				
1.5	5	103	257	5.8
1.5	3	62	154	9.7
3.0	5	103	257	11.7
3.0	3	62	154	19.4
Fuel at $5.00/gal				
1.5	5	103	514	2.9
1.5	3	62	309	4.9
3.0	5	103	514	5.8
3.0	3	62	309	9.7
Fuel at $10.00/gal				
1.5	5	103	1,028	1.5
1.5	3	62	617	2.4
3.0	5	103	1,028	2.9
3.0	3	62	617	4.9
Fuel at $20.00/gal				
1.5	5	103	2,057	0.7
1.5	3	62	1,234	1.2
3.0	5	103	2,057	1.5
3.0	3	62	1,234	2.4

mately $400 million (FY07$). In the KC-135R/T worst case, net savings become positive 24 years after starting the modification program. Only 217 of the 417 aircraft in the inventory are modified—the others are not modified because they are expected to be retired from the inventory before reaching the ends of their payback periods. Total net savings to the Air Force are approximately $36 million (FY07$).

As shown in Figure 4-3, the KC-10 best-case net savings become positive 8 years after starting the modification program. All 59 aircraft in the inventory are modified. Total net savings to the Air Force are approximately $221 million (FY07$). In the KC-10 worst case, net savings become positive 23 years after starting the modification program. Only 53 of the 59

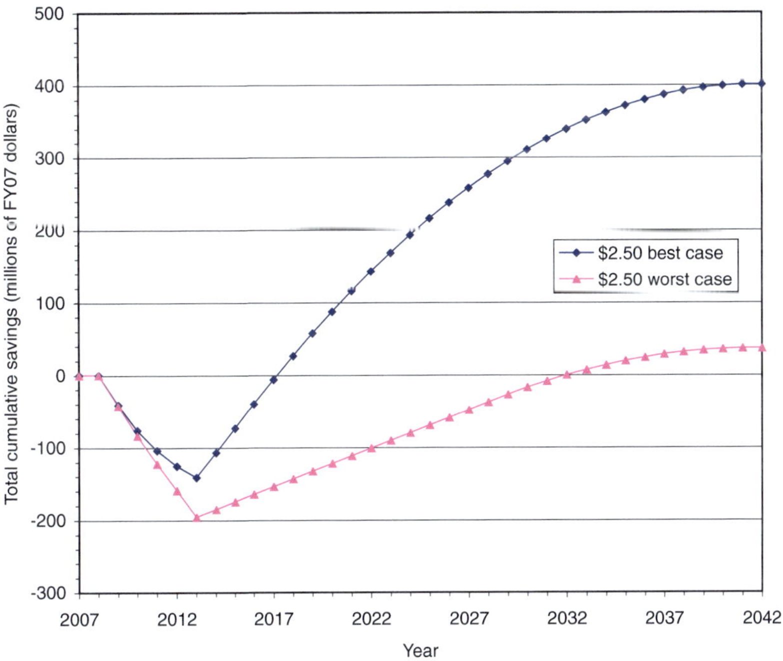

FIGURE 4-2 KC-135R/T estimated cumulative inventory-level net savings.

aircraft in the inventory are modified—the others are not modified because they are expected to be retired from the inventory before reaching the ends of their payback periods. Total net savings to the Air Force are approximately $12 million (FY07$).

Figure 4-4 illustrates how the cost of fuel affects net savings. The KC-135R/T worst-case payback periods are shown at fuel costs of $2.50, $5.00, $10.00, and $20.00 per gallon.

In constant dollars, when the cost of fuel is doubled, the payback period is cut in half. Total net savings to the Air Force rise significantly. The committee's analyses give only rough estimates of the costs and benefits of the modifications but, for reasonable projected values of the various factors, these rough estimates suggest that further analysis is warranted.

Finding: The committee's analysis for a broad range of fuel prices and with the data available to it on potential improvements in block fuel savings, modification cost estimates, operational parameters for the

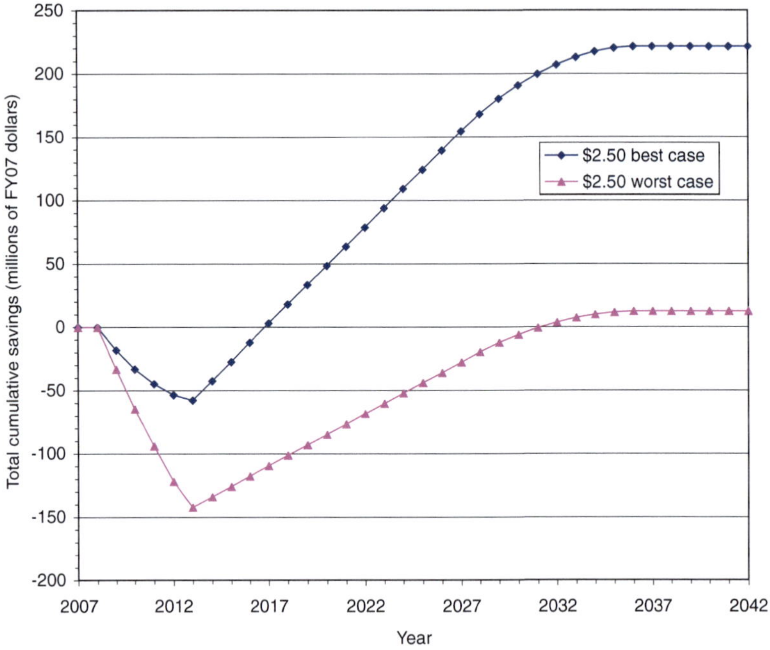

FIGURE 4-3 KC-10 estimated cumulative inventory-level net savings.

aircraft, and so forth indicates that wingtip modifications offer significant potential for improved fuel economy in certain Air Force aircraft, particularly the KC-135R/T and the KC-10.

Recommendation: The Air Force should initiate an engineering analysis with the original equipment manufacturers (OEMs) to determine (1) the extent and cost of modifications needed for the KC-135R/T and the KC-10 to enable installation of wingtip devices and (2) the fuel savings that could be achieved by this modification for each aircraft type. It should then perform an NPV analysis with these data to calculate the net savings. The Air Force should also analyze the C-5 and C-17 for potential wingtip modifications.

Once these analyses have been performed, more precise values for the modification costs and fuel savings will be known. The NPV calculations will give an idea of how long it takes to recover the investment. Note that

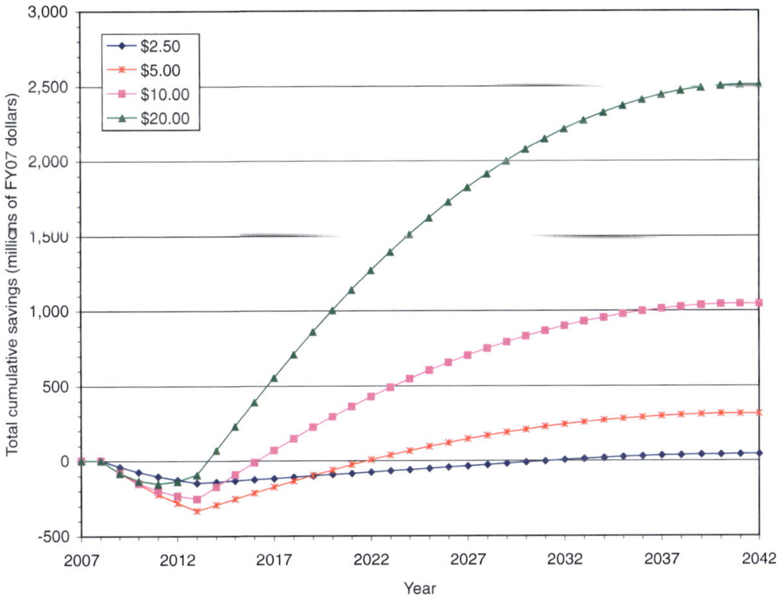

FIGURE 4-4 KC-135R/T effect of cost of fuel on payback period.

an important parameter in the NPV calculation is the price of fuel, which cannot be known in advance but instead must be hazarded. In any event, based on this preliminary analysis and the current price of fuel, these modifications are worthy of very serious consideration and analysis.

INVESTMENT STRATEGIES

The statement of task for this study asks for "investment strategies that the Air Force could implement with commercial partners to minimize Air Force capital investment and maximize investment return." Based on the analysis presented in this and earlier chapters, the committee proposes that the Air Force (1) follow through on its recommendation to initiate detailed engineering analysis in collaboration with the OEMs, (2) implement the modifications, if deemed cost effective, while the aircraft are in depot and in collaboration with industry, and (3) use innovative financing mechanisms as needed. The committee also suggests that the Air Force evaluate the fuel economy practices of commercial aircraft operators, some of which are described in Appendix B, and implement those that are applicable and not currently used by the Air Force. The strategies for investing in wingtip modifications are described in further detail below.

Performing Retrofit Studies and Implementing Modifications

Fuel economy has been a primary focus of commercial aircraft operators for a number of years. They have done an excellent job of working with the airframe manufacturers to perfect the aerodynamic design of aircraft to include wingtip modifications that will reduce drag, of implementing maintenance and operations procedures that save fuel, and of making fuel conservation a part of everyone's job and a factor in every decision. As a result, it is not surprising that this committee believes the aircraft with highest priority for further analysis are the KC-10 and the KC-135, two derivatives of commercial aircraft. The fact that these aircraft are commercial derivatives means that there is extensive commercial knowledge and experience to complement the military knowledge and experience. It also means that aerodynamic modifications have been examined more carefully and that more experienced engineers and maintenance personnel exist in the commercial industry than would be the case for military-unique aircraft such as the C-5, making the engineering analysis somewhat easier and increasing the availability of information. In any case, the feasibility and cost effectiveness of wingtip modifications on all of the aircraft should be worked out in partnership with the OEMs, whose knowledge of the aircraft structures and load distributions will be critical. In each case, the feasibility studies should be initiated as soon as possible. Then, a high priority should be given to funding the installation of wingtip modifications where they have been determined to be justified from a cost/benefit perspective. The sooner the modifications are incorporated, the sooner they will begin to pay back the initial investment and the less dependent the United States will be on foreign sources of fuel.

In addition, the KC-10 and the KC-135 constitute the aerial refueling capability of the Air Force and as such are force multipliers. As the fuel efficiency of these aircraft improves, they can either extend their range, carry more payload (i.e., offload more fuel to other aircraft), or do a combination of both things.

KC-10

In the case of the KC-10, winglet design work and testing have already been done on its commercial counterpart, the DC-10, as noted in Chapter 3. Although winglets were never incorporated on commercial DC-10 fleets, the knowledge gained from the engineering analysis, design work, and flight tests led to the installation of winglets on the MD-11. There is also the potential that commercial DC-10 operators such as FedEx could follow the Air Force lead and thus create a larger market for wingtip device modifications to the KC-10/DC-10s.

Should the decision be to proceed with such a modification, the committee suggests that the work be done while the aircraft are in normal scheduled overhaul. Since the KC-10 is maintained on contract with industry engineers who have intimate knowledge of commercial DC-10s, it is possible that wingtip modification could be added to the work specification with little or no added downtime or loss of operational availability.

KC-135R/T

Much of the same applies to the KC-135R/T aircraft fleet, except that unlike the KC-10, these aircraft are predominately maintained by Air Force personnel in-house. Also, as noted in Chapter 3, aerodynamic studies of wingtip modifications were done in the 1970s, and a test aircraft was modified with winglets and flight tested. Since the analysis and tests were done so many years ago and there are some uncertainties surrounding the condition of the KC-135 wings and their ability to handle the load increases from wingtip modifications, a sample of the fleet would have to be inspected. The best opportunity to do such an inspection or condition analysis is while the aircraft is in depot maintenance. Most depot overhauls of the KC-135s are performed by the Air Force at the Oklahoma City Air Logistics Center. During maintenance, paint is removed, engines are removed, and the aircraft are opened up for inspection of structural integrity, providing an excellent opportunity to take a careful look at the wings with minimal impact on depot flow. Like the KC-10s, these aircraft are critical to the operational commands, and every effort should be made not to increase scheduled downtime in the maintenance shops.

Should the modifications be justified, the committee believes the wingtips could be retrofitted while the aircraft are undergoing their 5-yearly depot maintenance. Rather than divert Air Force mechanics from other tasks, however, it might be wiser to partner with industry and have an experienced contract field team augment the Air Force workforce to accomplish the modification. This would minimize the training required and allow returning the aircraft to the operational forces in the shortest time. For the

KC-135R/T undergoing programmed depot maintenance at contractor facilities, the Air Force should consider adding wingtip modifications to the existing overhaul contract.

Other Aircraft

The next priority aircraft for consideration of wingtip modifications are the C-5 and the C-17. The same factors discussed in the investment strategies for the KC-10 and the KC-135R/T should be part of the planning process for these fleets as well.

Financing Mechanisms

Wingtip modification programs and other fuel economy investments are examples of long-term investments that may require a significant initial investment that provides returns over time. Securing financing for such long-term investments is always a challenge given the current military acquisition practices and congressional appropriation processes. In a previous report on engine fuel economy in military aircraft,[4] the NRC discussed innovative financing mechanisms that might be pursued. The statement of task for that study included a request to "develop implementation strategies to include conventional as well as innovative, acquisition, financing, and support concepts." The committee believes that three of the mechanisms discussed in that report—specifically, creating a line item in the defense budget, implementing an "energy savings performance contract" strategy, and competing airframe maintenance contracts—could be applicable in implementing wingtip modifications. Those mechanisms are discussed in some detail in the earlier report.

[4]NRC, 2007, *Improving the Efficiency of Engines for Large Nonfighter Aircraft*, Washington, D.C.: The National Academies Press.

Appendixes

Appendix A

Data Used in Net Present Value Analyses

The data values used in the net present value analyses discussed in Chapter 4 are shown in Tables A-1 and A-2.

TABLE A-1 KC-10 Data

Average individual aircraft flight hours per year		783
Average fuel consumption of an individual aircraft (lb/hr)		17,600
Weight of fuel (lb/gal)		6.7
Number of aircraft in inventory in 2007		59
Number of aircraft retired from inventory at start of year		
2028	6	
2029	6	
2030	6	
2031	6	
2032	6	
2033	6	
2034	6	
2035	6	
2036	6	
2037	5	

TABLE A-2 KC-135R/T Data

Average individual aircraft flight hours per year (hr/yr)		425
Average fuel consumption of an individual aircraft (lb/hr)		10,224.2
Weight of fuel (lb/gal)		6.7
Number of aircraft in inventory in 2007		417
Number of aircraft retired from inventory at start of year		
2018	16	
2019	17	
2020	17	
2021	16	
2022	17	
2023	17	
2024	16	
2025	17	
2026	17	
2027	16	
2028	17	
2029	17	
2030	16	
2031	17	
2032	17	
2033	16	
2034	17	
2035	17	
2036	16	
2037	17	
2038	17	
2039	16	
2040	17	
2041	17	
2042	17	

Appendix B

Additional Methods for Improving Fuel Consumption

The statement of task for this study focuses on the fuel economy of military aircraft and the potential of wingtip devices to reduce fuel consumption. However, wingtip devices are just one method for reducing fuel consumption. Other methods include making other aerodynamic modifications to the aircraft, improving engine efficiency, changing maintenance and operation practices, and improving weight management. Many of these strategies have already been adopted by the commercial airlines, which operate in an intensely competitive environment,[1] and others have been touched upon by several recent studies.[2] The committee believes it is important for these strategies to be considered, and while they were not the focus of this study, nor was the extent to which the Air Force may already be using some of them examined, some examples are discussed below for the reader's benefit.

Based on commercial experience, these other methods are expected to be relatively inexpensive, easy to implement, and could yield fuel consumption benefits comparable to wingtip devices. This appendix first explains

[1]Joseph C. Anselmo, 2004, "Airline fuel crisis," *Aviation Week & Space Technology* (December 6):54-56.

[2]Past studies on fuel conservation measures in the Air Force and at DOD include Defense Science Board (DSB), 2001, *More Capable Warfighting Through Reduced Fuel Burden*; Air Force Scientific Advisory Board, 2006, *Technology Options for Improved Air Vehicle Fuel Efficiency*; and NRC, 2007, *Improving the Efficiency of Engines for Large Nonfighter Aircraft*. Each of these studies included at least some discussion on current commercial practices.

some of the challenges experienced by commercial aircraft and then discusses other strategies for improved fuel efficiency. Since the preceding NRC report dealt with improving engine efficiency, an important determinant of fuel consumption, that strategy is not covered here.[3]

CHALLENGES

The aging and service use of commercial aircraft and jet engines take a toll, reducing aerodynamic and propulsion efficiency, as evidenced by increased fuel burn. As aircraft age and material wears, or suffers minor damage, fuel efficiency tends to decline because of external repairs, increased air leakage from the fuselage, weight gain from the entry of moisture and from years of modification programs, and engine deterioration. It is common for new commercial aircraft types to experience fuel burn increases over the specification (or "book") level of 2-4 percent within 4 years of entry into service. The regulatory agencies and internal technical organizations that certify continued airworthiness set the allowable in-service expansion of the original by tight manufacturing tolerances to accommodate the effects of normal wear and tear on commercial machinery.

Then, too, owners and operators of aircraft often push the performance limits of their equipment to achieve greater payload, range, endurance, or takeoff performance. Regardless of the specifications that prevailed when the aircraft were procured, political, regulatory, economic, or demographic influences open up prospects for new missions or markets that lie tauntingly just beyond the existing capabilities of existing in-service aircraft. Aircraft operators must then either seek new equipment with the required performance or attempt to improve the performance of existing equipment, through modification, to accommodate those new missions and markets. Specific strategies to take on these challenges are discussed below.

AERODYNAMICS

Lessons learned from the commercial airplane industry suggest that aerodynamic improvements using strategies other than wingtip modification are worth consideration for the Air Force's fleet of aircraft. Many of the its transport aircraft were designed in the early days of swept-wing trans-

[3]NRC, 2007, *Improving the Efficiency of Engines for Large Nonfighter Aircraft,* Washington, D.C.: The National Academies Press.

port design and do not take advantage of some more recent technological advancements, such as supercritical aft-loaded wings; low-interference, pod-mounted engine installations; reduced static stability; and digital designs with low excrescence drag.

Wing Modifications

A number of common performance improvements have been incorporated into the commercial fleet, both by the original equipment manufacturers (OEMs) and by third-party aircraft modification firms. Obviously, winglets are the most visible sign of this activity. Another common modification of earlier generation aircraft is re-rigging of the high-lift devices for cruise flight, creating a pseudo-aft-cambered wing. This has been done for the Boeing 727, for example. Another modification is the addition of a small, trailing-edge wedge on the lower surface of the wing. This creates some aft-camber and can also be used to change the span loading of the wing. That strategy was implemented on the MD-11 derivative of the DC-10 wing and is being studied for use on other aircraft. These trailing-edge modifications can be worth a reduction in fuel burn of up to 2.5 percent, depending on factors such as wing flexibility, trim drag characteristics, the original wing airfoil design, etc.[4]

Engine Installation

Pod-mounted engine installations of early-generation aircraft were crude by the standards of today, when high-powered computational fluid dynamics (CFD) methods have allowed very close coupling of engines with little or no interference drag. If a re-engine program is considered for a transport-category airplane, it is likely that a new engine installation can take advantage of this technology, resulting in a shorter pylon with less weight and wetted area and perhaps less interference drag as well. It is not likely that redesign of an existing engine installation to reduce drag or weight would pay off on its own, but if combined with a re-engine program, there could be a synergistic payoff of 1-2 percent.

---

[4]R.D. Gregg, R.W. Hoch, and P.A. Henne, 1989, "Application of divergent trailing-edge airfoil technology to the design of a derivative wing," SAE Technical Paper 892288, September; P.A. Henne and R.D. Gregg, 1991, "New airfoil concept," *AIAA Journal of Aircraft* 28(5):300-311.

Aerodynamic Cleanup

Aerodynamic cleanup programs are common, both for in-production and in-service airplanes. This would include redesign of excrescences, such as door seals, high-lift system seals, rigging, antenna installations, protruding fasteners, air inlets and exhausts for external air exchange systems, and so on. It also might include redesign of aerodynamic fairings, including flap support fairings, wing-to-body fairings, and the like. Up to 4 percent of airplane drag has been saved on commercial aircraft, some having cleanup programs and others not. As an example, the MD-11 had a Cruise Performance Improvement Program, which resulted in approximately a 4 percent improvement to the fuel burn efficiency of the modified aircraft.[5] Further investigation would be required to determine if any of these redesigned items, which were above and beyond the basic improvements made to the original MD-11 design by incorporation of the winglets and trailing-edge wedges, are applicable to the KC-10/DC-10 family.

MAINTENANCE AND OPERATIONS

The mechanical condition of an aircraft and the means by which it is operated are critical for maintaining original performance design characteristics and objectives. As stated earlier, commercial aircraft typically exhibit fuel burn 2-4 percent above the book value within 4 or so years of entering service. Airline experience demonstrates that it is difficult to determine the relative contribution of the airframe and the engine to this fuel burn deterioration. Over the years, the airlines and commercial aircraft and engine manufacturers have developed comprehensive maintenance and operational procedures to return aircraft to their certified fuel-burn performance. Collectively, these efforts can improve fuel burn by 1-3 percent. These procedures are effective and relatively easy to implement. Where these procedures make operational sense and are not currently used by the Air Force, military managers should consider implementing the practices that have merit.

---

[5]Robb Gregg, Senior Manager for Aircraft Programs, Boeing Phantom Works, "Drag improvement: A study of the DC-10/MD-11/C-17 winglet programs," Presentation to the committee on December 13, 2006.

Maintenance

Initial efforts to improve performance generally rest with an attempt to regain the original tolerances and material conditions for in-service aircraft. These efforts are generally accomplished according to priorities that are jointly developed with the OEM. Based on individual airline operating experience, these maintenance activities or fine-tuning exercises to return an aircraft as close as practical to its original material condition and configuration will frequently reduce fuel burn by 1-3 percent (or possibly more).

Effective maintenance programs require a comprehensive knowledge of the mechanical condition of the aircraft and its systems and the conditions that cause mechanical malfunctions. They require, as well, a detailed accounting of the maintenance actions conducted and the resulting effect on the malfunction. Most important, program success requires the development of measures and standards for efficient operation of the equipment.

Maintenance programs must be developed to take into account some of the systems and elements that, if not operating properly, can have a major negative impact on fuel burn:[6]

- *Air data.* Air data generally refer to the aircraft's pitot-static system, which gives crew and system a reference for airspeed, altitude, and vertical velocity. Air data refer as well to some engine instrumentation such as engine pressure ratio, which gives crew and systems proper engine power information. Proper maintenance of these systems is essential to assure that the aircraft is operating at the airspeed/Mach number, altitude, and power that give the most efficient fuel burn. In addition, improper power setting can result in asymmetric thrust, which must be compensated for by trimming the control surfaces, increasing drag. The commercial industry recently went through an accuracy improvement in air data systems to support the worldwide Reduced Vertical Separation Minima program. This revealed system deficiencies that have resulted in system improvements to assure optimum operational and fuel burn performance. The technology is now available that would allow collecting more accurate airspeed data.

---

[6]These are also discussed in *Improving the Efficiency of Engines for Large Nonfighter Aircraft*, Washington, D.C.: The National Academies Press, 2007. That report also discusses improvements to the maintenance programs for engines when they are in depot (rather than on-wing). That discussion is omitted here.

- *Pneumatics.* Pneumatic leakage through door cutouts, improper sealing, airframe damage, and fuselage attach fittings adversely affects fuel burn in two ways: (1) extra fuel is consumed because the air-cycle machines must work harder to compensate for the leakage and (2) the leakage of air from the fuselage disrupts the airflow around the aircraft, resulting in increased drag. Close monitoring of the airframe and engine pneumatic systems is encouraged to maintain optimum fuel burn.
- *Seals.* It is essential to assure that the aerodynamic seals between the lower and upper wing are in good condition, especially on the leading edges.
- *Flight controls.* Flight controls must be properly rigged. Floating spoilers, flaps that are not properly seated, and ailerons not properly rigged can all have a very large impact on fuel burn. Large surfaces such as rudders are especially critical and adversely impact fuel burn if out of rig or trimmed to offset asymmetric thrust conditions.
- *Fuel indicators.* To assure the best flight profiles for fuel efficiency, it is essential to have accurate references for fuel quantity and fuel flow. In order to achieve this objective it is essential that fuel quantity probes and indicating systems as well as flowmeters be calibrated periodically.
- *Engine performance.* Over time, the wear on engine blades adversely affects the gas path of turbine engines. The earliest sign of these effects is commonly the loss of exhaust gas temperature (EGT) margins. This loss is typically between 5°C and 7°C of EGT per 1,000 hours of flight time and ultimately impacts takeoff performance, especially at hot or high-altitude airports with relatively short runways. This deterioration can be mitigated by a rigorous on-wing engine wash program that initially returns between 5°C and 10°C of EGT. However, as the engine continues to deteriorate over time, this effect decreases as well.
- *Housekeeping.* Simple housekeeping actions can have benefits, such as maintaining leading edges so that they are clean and free of excessive dents, making sure the pitot-static lines are free of obstructions, and assuring the proper calibration and functioning of systems to measure air mass temperature. The removal of fittings and materials remaining from past modifications or temporary accoutrements that add unnecessary weight to the airframe is also important. The importance of reducing unnecessary weight is discussed elsewhere in this appendix.

Operations

A number of operational procedures and practices have been developed by the air transport industry to reduce fuel consumption. Their effectiveness is dependent on (1) the commitment of management and flight crews to their use and (2) standardization in their application throughout all functions of the organization.

The following elements are fundamental to controlling excessive fuel burn. They are well known by all aircraft operators. To the extent that they are effectively managed to affect fuel burn depends on how ingrained they are into the thought processes of individual flight, maintenance, planning, and configuration control personnel—in other words, how well they are accepted into the culture of the organization.

Fuel Burn Tracking

Most airlines have strict fuel burn reduction plans that track individual aircraft and flight crews to isolate equipment or operational factors that contribute to excessive fuel burn. The plans, which are frequently developed in conjunction with the aircraft manufacturer, include the following:

- Develop flight-phase operational configurations and profiles—that is, takeoff and climb to cruise, cruise, descent/land profiles—to provide the optimum airspeed and power setting for targeted fuel burn and flight performance at the given gross weight and altitude of the aircraft.
- Report periodically while in flight on fuel burn, power settings, airspeed, and altitude.
- Determine block fuel use for specific aircraft and flight crews.

Continuous monitoring of cruise performance can give aircraft operators the information they need to decide how and where to save fuel. Such monitoring allows the operators to do the following:

- Adjust the baseline performance levels they use for flight planning so that the correct amount of fuel is loaded on each and every flight.
- Increase flight crew confidence in flight plans and possibly decrease the amount of discretionary fuel requested.
- Identify airplanes that burn a lot of fuel for possible corrective actions.

- Match the airplanes and engines that perform best with respect to fuel burn to fly the longest range/endurance missions.

If a specific aircraft is flagged as having excessive fuel burn, maintenance action is initiated to determine, and correct, the cause of that unnecessary burn (the preceding section on maintenance gives details). Airframe and engine manufacturers may be called on to assist if the corrective actions are not readily identifiable.

If a particular flight crew, or flight crew member, consistently exceeds average block fuel usage for specific flight segments, the situation may be addressed with appropriate training. Wherever possible, the flight crew should assure that its fuel burn practices comply with the following guidelines:

- Use the manufacturer-recommended fuel burn procedures for wing tanks as appropriate to maintain wing structural integrity and stiffness.
- Maintain lateral balance during fuel burn.
- Maintain aft center of gravity (CG) with fuel burn.

Trim

One of the main reasons specific aircraft and/or flight crew members have excessive fuel burn is improper trim, which can come from a sub-optimal performance indicating system, fuel quantity system, or flight control rig or from poor flight crew performance. Airline experience has demonstrated that even pilots with thousands of flying hours and years of experience in the cockpit can fail to trim aircraft properly.

A number of priorities must be observed to properly trim an aircraft. When the mission requires predominant use of the autopilot, the flight crew should assure that the aircraft is trimmed properly prior to connecting the autopilot and should then disconnect the autopilot periodically to retrim as necessary. Proper aircraft trim is achieved by the following means:

- Maintain lateral balance during fuel burn.
- Fly the aircraft manually to maintain straight and level flight.
- Balance the thrust using all of the engine performance indicators.
- Trim the elevator to eliminate elevator control force and maintain level flight.

- Trim the rudder to eliminate rudder control force and sideslip/turning flight.
- Trim the aileron to eliminate control force.
- Verify control displacements (spoilers, ailerons, and rudder within manufacturer/service limits) for potential maintenance action (rigging).

As mentioned in the maintenance section, it is important to verify control displacements (spoilers, ailerons, and rudder should be monitored within the manufacture's service limits) for potential maintenance action (rigging). Also, it is obvious that failure to calibrate flight and performance instrumentation will prevent the flight crew from trimming the aircraft properly.

Ground Operations

Standard procedures exist for ground operations as well to minimize unnecessary use of engine power and the auxiliary power unit (APU). The following exemplify such procedures:

- Single-engine taxi is used for two-engine aircraft, and one- or two-engine shut-down taxi for three-engine and four-engine aircraft, whenever the airport and operational conditions and configurations allow.
- Engines are not started until the appropriate time in the departure sequence.
- The APU is not used until required for engine start or postflight operations unless external conditions require it (high temperatures, absence of ground power, etc.).

WEIGHT MANAGEMENT

The main goal of aircraft manufacturers is to design their aircraft to carry out the intended mission with the best possible performance. A common objective relative to that goal is to eliminate as much unnecessary weight and material as possible. This is true because every added pound of weight eats into aircraft performance margins by feeding the twin detriments of unnecessary fuel burn and reduced payload. Two facts are certain to apply to almost every commercial or military aircraft: (1) The basic aircraft empty weight will increase over the life of the aircraft (to the detriment of payload capability and fuel burn performance) and (2) mission demands will grow to push the operational limits of the aircraft.

To address these realities, aircraft operators must work diligently and continuously to determine and control the actual weight and balance of their aircraft. This is accomplished by programs that allow the following:

- Periodic and accurate determination of individual aircraft weight and balance (CG).
- Controlling aircraft modification programs to minimize weight increases and maintaining allowable CG aft to reduce drag.
- Maintaining the external condition of the aircraft to maintain aerodynamic efficiency and minimize drag—for example, assure that dirt and other external contaminants such as grease build due to cleaning lubricants and the like do not add weight or affect the aerodynamics.
- Calibrating flight and performance instrumentation to assure proper criteria for weight, flight conditions, and performance.

The following are examples of additional and relatively simple actions that can be taken to reduce fuel consumption:

- Establish a baseline of equipment and material routinely carried on the aircraft (pallets, tools, etc.). Obtain fleet aircraft weight samples to determine the spread in actual weights, including weighing some operational aircraft ready to go out on a mission and some empty aircraft. Weigh all the equipment that is put on aircraft, such as repair kits. Use actual rather than estimated weights for cargo. Load all materials so as to maintain the maximum allowable (or practical) aft CG.
- Revise operational practices to reduce unnecessary weight. For training and operational flights, eliminate any equipment that is not essential to the mission. Do not carry excess fuel since its weight increases fuel consumption. Review the need to carry remote station tools and equipment and accurately account for the weight of necessary tools and equipment. Weigh all cargo to verify that registered weights are accurate.
- Revise maintenance practices to reduce unnecessary weight. Ensure aircraft are clean and not carrying water, trash, or dirt in cavity and

> swamp areas. Check insulation blankets for condensation which can increase the weight of the blankets significantly—by, for example, more than 1,000 lb in the case of 707 blankets. Consider lighter weight replacement materials for nonstructural items such as floor panels (floors in KC-135s, for example, are plywood). Create a weight maintenance czar to keep aircraft weight as stable as possible over time.

The commercial airline industry has also employed changes when designing new aircraft to improve CG management. Newer designs, such as the Boeing 777 and 787 and the MD-11, have used stability augmentation to allow smaller tail surfaces and to shift the CG aft, reducing trim drag. For an existing aircraft, it is probably not practical to change the design to improve stability or allow smaller tail surfaces. But, as mentioned above, by paying careful attention to payload loading position, an aircraft can be routinely flown near its aft CG limit, often saving a percent or more in trim drag. Commercial airlines have automated their loading processes to make aft loading more routine.

Appendix C

Presentations to the Committee

MEETING 1
DECEMBER 13-14, 2006
WASHINGTON, D.C.

Wingtip Devices: What They Do and How They Do It
Doug McLean, Boeing Technical Fellow
Boeing Commercial Airplanes

Overview of Winglets on Boeing Commercial Airplanes
Mark Goldhammer, Committee Member
Committee on Assessment of Aircraft Winglets for Large Aircraft Fuel Efficiency

American Airlines Winglets
John Novelli, Director, Operations Engineering and Optimization
American Airlines

Drag Improvement: A Study of the DC-10/MD-11/C-17 Winglet Programs
Robb Gregg III, Senior Manager
Boeing Phantom Works

C-5 and C-130 Discussion
Lane Ito, Advanced Development Programs
Lockheed Martin Aeronautics Company

Past Winglet Studies: Air Force Scientific Advisory Board Fuel Efficiency Study
Ilan Kroo, Committee Member
Committee on Assessment of Aircraft Winglets for Large Aircraft Fuel Efficiency

Winglets Experience at Southwest (teleconference)
Jim Sokol, Vice President of Maintenance and Engineering
Southwest Airlines

Appendix D

Biographical Sketches of Committee Members

Kenneth E. Eickmann, *Chair,* retired from the Air Force after a 31-year career in which his last assignment was commander of the Aeronautical Systems Center within the Air Force Materiel Command at Wright-Patterson Air Force Base, Ohio. In that capacity he led the Air Force's center of excellence for research, development, and acquisition of aircraft aeronautical equipment and munitions. His leadership accomplishments also include having led the federal rescue and recovery efforts following the 1995 bombing of Oklahoma City's Alfred P. Murrah Building. More recently, he served as the director of the Construction Industry Institute (CII) at the University of Texas (UT) at Austin. CII, a nonprofit research institute, is the principal national forum for the multitrillion-dollar-a-year construction industry. Gen. Eickmann earned a B.S. in mechanical engineering from UT Austin in 1967 and an M.S. in systems engineering from the Air Force Institute of Technology in 1968. He is also a graduate of the University of Michigan Executive Business Program and the John F. Kennedy School of Government at Harvard University. Gen. Eickmann is currently a member of NRC's Air Force Studies Board and was chair of the NRC's Committee on Analysis of Air Force Engine Efficiency Improvement Options for Large Non-fighter Aircraft.

Natalie W. Crawford (NAE), *Vice Chair,* is senior fellow and former vice president and director of Project Air Force (PAF) at the RAND Corporation. Since joining RAND in 1964 as a member of the Engineering Sciences and Aeronautical and Astronautics departments, she has held a wide variety of research and administrative posts. She has led PAF research on aircraft survivability, conventional standoff weapons, tactical aircraft, electronic combat, and integrated avionics for the advanced tactical fighter. As director of PAF's Theater Force Employment Program, Mrs. Crawford formed a team of analysts to compile and edit Desert Storm air campaign data, leading to the first usable databases for analysis of that campaign. While associate director of PAF (1995-1997), she was in charge of a comprehensive, multidisciplinary analysis of the roles and capabilities of the Air Force in the 21st century. Then, as director of PAF, Mrs. Crawford oversaw all research conducted at RAND for the U.S. Air Force. In FY00, at the request of the Air Force chief of staff, she led a major review of requirements, acquisition, operations, and sustainment of Air Force electronic warfare programs and systems, culminating in a four-star summit chaired by the chief of staff. She has been a member of the Air Force Scientific Advisory Board since 1988, serving as its vice chairman in 1990 and 1991 and co-chairman from 1996 to 1999. To develop insight and understanding in her research, she has flown missions in several Air Force aircraft. In 2003, she was awarded the Vance R. Wanner Memorial Award from the Military Operations Research Society. She received a B.A. in mathematics from the University of California at Los Angeles, where she also pursued graduate studies in engineering. Mrs. Crawford was also vice chair of the NRC's Committee on Analysis of Air Force Engine Efficiency Improvement Options for Large Non-fighter Aircraft.

Mark I. Goldhammer is the chief engineer for the Product Development Airplane Performance organization at Boeing Commercial Airplanes. In this position, he has functional oversight of the airplane performance disciplines assigned to the 787 and product development, including responsibilities for the 747-8, derivative and new airplane product development, advanced concepts, and competitive analysis. Mr. Goldhammer joined Boeing Commercial Airplanes in early 1977 and has worked on a variety of product development studies in high-lift aerodynamic design methods, transonic wing design, wind tunnel testing, and other aerodynamic design issues. He held positions as engineer, lead engineer, and manager of aerodynamics engineering on the 777 and was responsible for the aerodynamic configuration design from preliminary design through flight testing and certification. Mr. Goldhammer has also held managerial responsibilities for the aerodynamic configuration development of a rewinged/stretched derivative of the 747 and the certification of the 737-700C and the 737-900. He also represented the 737-NG on Boeing safety review boards and was instrumental in implementing lean principles to the delivery certification process for the 737-NG program. Prior to Boeing, Mr. Goldhammer began his career at the Douglas Aircraft Company. He received a B.S. in aeronautical engineering from Rensselaer Polytechnic Institute and an M.S. in aeronautical engineering from the University of Southern California. He is also a licensed private pilot.

Stephen Justice is concept exploration and development manager within Lockheed Martin's advanced development programs, also known as the Skunk Works, with responsibility for generating and developing new project ideas. Mr. Justice joined Lockheed in 1984 and held roles of increasing responsibility on programs that included the F-117 Nighthawk Stealth Fighter, YF-22 Stealth Air Superiority Fighter, and numerous classified programs. His aeronautical engineering experience ranges from conceptual design to preliminary design, detail design, fabrication liaison, flight test, design leadership, and program management. He has a B.S. in aerospace engineering from the Georgia Institute of Technology and has two awarded patents and five classified patent disclosures. Prior to Lockheed, Mr. Justice began his career in defense aerospace as a structural designer in Texas with General Dynamics' Fort Worth Division. In 2005, he received the LM Aeronautics Company AeroStar award and corporate NOVA award for leadership. Mr. Justice also is an instructor for Lockheed Martin Technical Institute in aircraft configuration development, structural design, systems design, and low observables (stealth) technology integration and is a licensed pilot.

Clyde Kizer retired in 2004 from Airbus Industries of North America as president of customer service. In that capacity, he had total customer services responsibilities for all Airbus aircraft operating in North America and spares and training responsibilities for all Airbus operators in the Western Hemisphere. Mr. Kizer's 12-year tenure with Airbus saw explosive growth for that company in North America, going from 98 Airbus aircraft of all types in North America to 980. Prior to Airbus, Mr. Kizer served as senior vice president of operations for Midway Airlines; vice president of engineering and maintenance at the Air Transport Association; and vice president of engineering at United. Mr. Kizer also served for 23 years as a Navy operational and experimental test pilot and flew 14 years as an engineering test pilot for United. He earned a degree in biochemistry from Eastern Michigan University in 1960. Mr. Kizer was also a member of NRC's Committee on Analysis of Air Force Engine Efficiency Improvement Options for Large Non-fighter Aircraft.

Ilan Kroo (NAE) is a professor in the Department of Aeronautics and Astronautics at Stanford University, where he also received a Ph.D. Prior to joining the faculty at Stanford in 1985, he worked in the Advanced Aerodynamics Concepts Branch at NASA's Ames Research Center for 4 years. His research in aerodynamics and multidisciplinary design optimization includes the study of innovative airplane concepts. He participated in the design of unmanned aerial vehicles (UAVs) modeled on (extinct) pterosaurs (flying reptiles), America's Cup sailboats, and high-speed research aircraft. In addition to his research and teaching interest, he is director of a small software company and is an advanced cross-country hang glider pilot. He is a fellow of the American Institute of Aeronautics and Astronautics. Dr. Kroo was elected to the National Academy of Engineering for new concepts in aircraft design methodology and for the design and development of the SWIFT sailplane.

Eli Reshotko (NAE) is Kent H. Smith Professor Emeritus of Engineering at Case Western Reserve University and currently resides in Denver. Dr. Reshotko joined the faculty at Case Western in 1964 and prior to that worked at NASA-Lewis Flight Propulsion Laboratory (now NASA-Glenn Research Center). Dr. Reshotko graduated from the California Institute of Technology with a Ph.D. in aeronautics and physics, and his expertise includes viscous effects in external and internal aerodynamics. He is a fellow of the following societies: American Institute of Aeronautics and Astronautics, the American Society of Mechanical Engineers, the American Physical Society, and the American Academy of Mechanics, of which he has served as president. He is coauthor of over 100 publications and is affiliated with many task forces, committees, and governing boards, many of which he served as chair. Dr. Reshotko was also a member of the NRC's Committee on Analysis of Air Force Engine Efficiency Improvement Options for Large Non-fighter Aircraft.

Raymond Valeika retired from Delta as senior vice president for technical operations (TechOps), where he directed a worldwide maintenance and engineering staff of more than 10,000 professionals, for a fleet of nearly

600 aircraft. Currently, he is an independent consultant advising major companies in aviation matters and an internationally recognized senior airline operations executive with over 40 years of experience managing large airline maintenance operations. Through his leadership and focus on continuous improvement of the human processes in aviation maintenance, Delta TechOps consistently rated at the top of the industry for performance benchmarks in the areas of safety, quality, productivity, and reliability. Prior to Delta, he held senior executive positions with Pan Am and Continental Airlines. In 1996, Mr. Valeika was honored with the Air Transport Association's Nuts & Bolts award, which recognized his leadership in the aviation industry. In October 1999, Mr. Valeika received the Marvin Whitlock Award from the Society of Automotive Engineers for his accomplishments and long-term leadership within the aeronautical engineering and commercial aviation industries. Most recently, the Aviation Week Group honored him with a lifetime achievement award. He is also a member of NRC's Aeronautics and Space Engineering Board. He graduated from St. Louis University with a degree in aeronautical engineering in 1964. Mr. Valeika was also a member of NRC's Committee on Analysis of Air Force Engine Efficiency Improvement Options for Large Non-fighter Aircraft.